Understanding and Using the Controller Area Network Communication Protocol

Marco Di Natale • Haibo Zeng • Paolo Giusto
Arkadeb Ghosal

Understanding and Using the Controller Area Network Communication Protocol

Theory and Practice

Springer

Marco Di Natale
Scuola Superiore S. Anna
RETIS Lab.
Pisa 56124, Italy

Paolo Giusto
General Motors Corporation
Palo Alto, CA 94306, USA

Haibo Zeng
McGill University
Montreal, Quebec, Canada H3A 2A7

Arkadeb Ghosal
National Instruments
Berkeley, CA 94704, USA

ISBN 978-1-4614-0313-5 e-ISBN 978-1-4614-0314-2
DOI 10.1007/978-1-4614-0314-2
Springer New York Dordrecht Heidelberg London

Library of Congress Control Number: 2011944254

Printed on acid-free paper

Springer is part of Springer Science+Business Media (www.springer.com)

Preface

This book is the result of several years of study and practical experience in the design and analysis of communication systems based on the Controller Area Network (CAN) standard. CAN is a multicast-based communication protocol characterized by the deterministic resolution of the contention, low cost, and simple implementation. The CAN [16] was developed in the mid 1980s by Bosch GmbH, to provide a cost-effective communications bus for automotive applications. Today it is widely used also in factory and plant controls, in robotics, medical devices, and also in some avionics systems.

Controller Area Network is a broadcast digital bus designed to operate at speeds from 20 kbit/s to 1 Mbit/s, standardized as ISO/DIS 11898 [6] for high speed applications (500 kbit/s) and ISO 11519-2 [7] for lower speed applications (up to 125 kbit/s). The transmission rate depends on the bus length and transceiver speed. CAN is an attractive solution for embedded control systems because of its low cost, light protocol management, the deterministic resolution of the contention, and the built-in features for error detection and retransmission. Controllers supporting the CAN communication standard, as well as sensors and actuators that are manufactured for communicating data over CAN, are today widely available. CAN networks are successfully replacing point-to-point connections in many application domains.

Commercial and open source implementation of CAN drivers and middleware software is today available from several sources, and support for CAN is included in automotive standards, including OSEKCom and AUTOSAR. The standard has been developed with the objective of time determinism and support for reliable communication. With respect to these properties, it has been widely studied by academia and industry, and methods and tools have been developed for predicting the time and reliability characteristics of messages.

The CAN standard has originally been proposed for application to automotive systems, but with time emerged as a quite appropriate solution for other control systems as well. This book tries a general approach to the subject, but in many places it refers to automotive standards and systems. Automotive architectures and functions are also often used as examples and benchmarks. We hope this is not a problem for the reader and hopefully, the generalization of the proposed approaches and methods to other domains should not be too hard. On the other side, automotive systems are today an extremely interesting context for whoever is interested in complex and distributed embedded systems (including of course those using the CAN bus). They offer complex architectures with multiple buses and nodes with gateways, real time and reliability constraints and all the challenges that come with a high bus utilization and the demand for increased functional complexity.

This book attempts at providing an encompassing view on the study and use of the CAN bus, with references to theory and analysis methods, and a description of the issues in the practical implementation of the communication stack for CAN and the implications of design choices at all levels, from the selection of the controller, to SW development and architecture design. We believe such an approach may be of advantage to those interested in the use of CAN, from students of embedded system courses, to researchers, architecture designers, system developers, and all practitioners that are interested in the deployment and use of a CAN network and its nodes.

As such, the book attempts at covering all aspects of the design and analysis of a CAN communication system. Chapter 1 contains a short summary of the standard, with emphasis on the bus access protocol and on the protocol features that are related to or affect the reliability of the communication. Chapter 2 describes the functionality, the design, the implementation options, and the management policies of the hardware controllers and software layers in CAN communication architectures. Chapter 3 focuses on the worst case time analysis of the message response times or latencies. Chapters 4 and 5 presents the stochastic and statistical timing analyses. Chapter 6 addresses reliability issues. Chapter 7 deals with the analysis of message traces. Chapter 8 describes commercial tools for configuring, analyzing and calibrating a CAN communication system. Chapter 9 contains a summary of the main transport level and application-level protocols that are based on CAN.

Pisa, Italy — Marco Di Natale
Montreal, QC, Canada — Haibo Zeng
Palo Alto, CA, USA — Paolo Giusto
Berkeley, CA, USA — Arkadeb Ghosal

Contents

List of Figures

List of Tables

Chapter 1
The CAN 2.0b Standard

This chapter introduces version 2.0b of the CAN Standard. This introduction is an excerpt of the main features of the protocol as described in the official Bosch specification document [16]. For more details, the reader should check the free official specification document available on-line, along with the other references provided throughout this chapter.

The CAN network protocol has been defined to provide deterministic communication in complex distributed systems with the following features and capabilities:

- Message priority assignment and guaranteed maximum latencies.
- Multicast communication with bit-oriented synchronization.
- System-wide data consistency.
- Bus multimaster access.
- Error detection and signaling with automatic retransmission of corrupted messages.
- Detection of permanent failures in nodes, and automatic switch-off to isolate faulty node.

In the context of the ISO/OSI reference model, the original CAN specification, developed by Robert Bosch GmbH, covers only the *Physical* and *Data link* layers. Later, ISO provided its own specification of the CAN protocol, with additional details for the implementation of the physical layer.

Generally speaking, the purpose of the Physical Layer is to define the encoding of bits into (electrical or electromagnetic) signals with defined physical characteristics. The signals are transmitted over wired or wireless links from one node to another. In the Bosch CAN standard, however, the description is limited to the definition of the bit timing, bit encoding, and synchronization. The specification of the physical transmission medium including the required (current/voltage) signal levels, the connectors, and other characteristics that are necessary for the definition of the driver/receiver stages and the physical wiring is not covered. Other reference documents and implementations have filled this gap, providing solutions for the practical implementation of the protocol.

M. Di Natale et al., *Understanding and Using the Controller Area Network Communication Protocol: Theory and Practice*, DOI 10.1007/978-1-4614-0314-2_1,

The Data-link layer consists of the Logical Link Control (LLC) and Medium Access Control (MAC) sublayers. The LLC sublayer provides all the services for the transmission of a stream of bits from a source to a destination. In particular, it defines:

- Services for data transfer and remote data request.
- Conditions upon which received messages should be accepted, including message filtering.
- Mechanisms for recovery management and flow management (overload notification).

The MAC sublayer is considered as the kernel of the CAN protocol. The MAC sublayer is responsible for message framing, arbitration of the communication medium, acknowledgment management, and error detection and signaling. For the purpose of fault containment and additional reliability, the MAC operations are supervised by a controller entity monitoring the error status and limiting the operations of a node if a possible permanent failure is detected.

The following sections provide more details into each sub-layer, including requirements and operations.

1.1 Physical Layer

As stated in the introduction, the Bosch CAN standard defines bit encoding, timing and synchronization, included in the Physical Signaling (PS) portion of the ISO-OSI physical layer. The standard does not cover other issues related to the physical layer, including the types of cables and connectors that should be used for communicating over a CAN network and the ranges of voltages and currents that are considered as acceptable for the operations. In the OSI terminology, the Physical Medium Attachment (PMA) and Medium Dependent Interface (MDI) are the two parts of the physical layer which are not defined by the original standard.

1.1.1 Bit Timing

The signal type is digital with *Non Return to Zero* (NRZ) bit encoding. The use of NRZ encoding ensures a minimum number of transitions and high resilience to external disturbance. The two bits are encoded in physical medium states defined as "recessive" and "dominant." (0 is typically assumed to be associated with the "dominant" state). The protocol allows multi-master access to the bus. At the lowest level, if multiple masters try to drive the bus state at the same time, the "dominant" configuration always prevails upon the "recessive."

Nodes are requested to be synchronized on the bit edges so that every node agrees on the value of the bit currently transmitted on the bus. To achieve synchronization, each node implements a protocol that keeps the receiver bit rate aligned with the actual rate of the transmitted bits. The synchronization protocol uses transition edges to resynchronize nodes. Hence, long sequences without bit transitions should be avoided to ensure limited drift among the node bit clocks. This is the reason why the protocol employs the so-called "bit stuffing" or "bit padding" technique, which forces a complemented bit in a transmission sequence after 5 bits of the same type. Stuffing bits are automatically inserted by the transmission node and removed at the receiving side before processing the frame contents.

Synchronous bit transmission enables the CAN arbitration protocol and simplifies data-flow management, but also requires a sophisticated synchronization protocol. Bit synchronization is performed first upon the reception of the start bit available with each asynchronous transmission. Later, to enable the receiver(s) to correctly read the message content, continuous resynchronization is required. Other features of the protocol influence the definition of the bit timing. Bus arbitration, message acknowledgement and error signalling are based on the capability of the nodes to change the status of a transmitted bit from recessive to dominant. Since the bus is shared, all other nodes in the network are informed of the change in the bit status before the bit transmission ends. Therefore, the bit (transmission) time must be at least large enough to accommodate the signal propagation from any sender to any receiver and back to the sender.

The bit time needs to account for a propagation delay that includes the signal propagation delay on the bus as well as delays caused by the electronic circuitry of the transmitting and receiving nodes. In practice, this means that the signal propagation is determined by the two nodes within the system that are farthest apart from each other as the bit is broadcasted to all nodes in the system (Fig. 1.1).

The leading bit edge from the transmitting node (node A in Fig. 1.1 reaches node B after the signal propagates all the way from the two nodes. At this point, B can change its value from recessive to dominant, but the new value will not reach A until the transition from recessive to dominant propagates across the entire bus length from B back to A. Only then can node A safely determine whether the signal level it wrote on the bus is the actual stable level for the bus at the bit sampling time, or whether it has been replaced (in case it was recessive) by a dominant level superimposed by another node.

Considering the synchronization protocol and the need that all nodes agree on the bit value, the nominal bit time (reciprocal of the bit rate or bus speed) is defined as composed of four segments (Fig. 1.2, segment names are all upper case in accordance with the notation in the original specification.)

- *Synchronization segment (SYNC_SEG)* This is a reference interval, used for synchronization purposes. The leading edge of a bit is expected to lie within this segment.
- *Propagation segment (PROP_SEG)* This part of the bit time is used to compensate for the (physical) propagation delays within the network. It is twice the sum of the signal propagation time on the bus line, plus the input comparator delay, and the output driver delay.

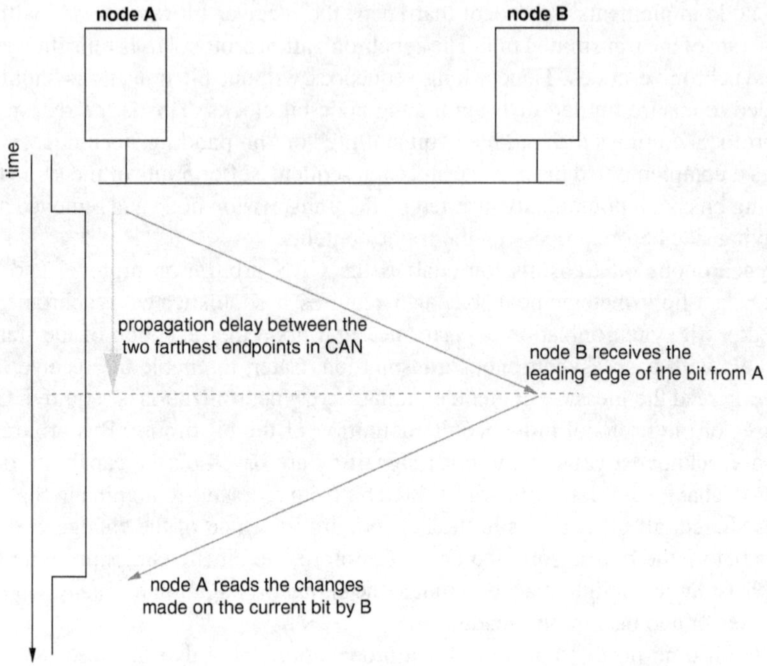

Fig. 1.1 Bit propagation delay

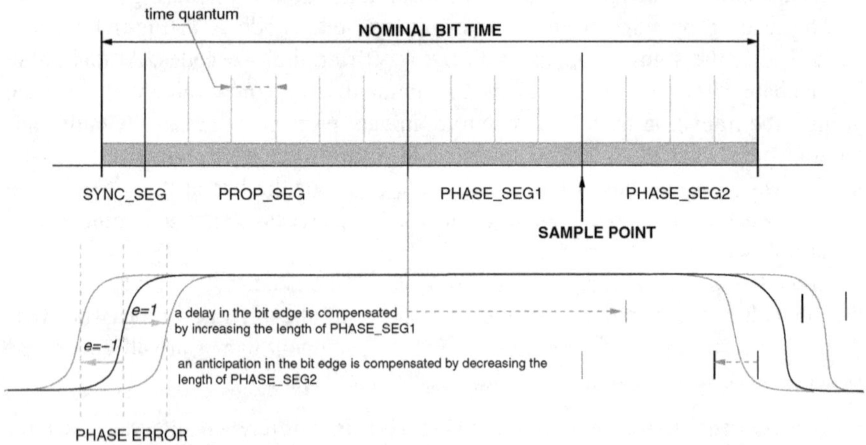

Fig. 1.2 Definition of the bit time and synchronization

- *Phase segments (PHASE_SEG1 and PHASE_SEG2)* These phase segments are time buffers used to compensate for phase errors in the position of the bit edge. These segments can be lengthened or shortened to resynchronize the position of SYNCH_SEG with respect to the following bit edge.
- *Sample point (SAMPLE_POINT)* The sample point is the point of time at which the bus level is read and interpreted as the value of that respective bit. The quantity *information processing time* is defined as the time required to convert the electrical state of the bus, as read at the SAMPLE_POINT into the corresponding bit value.

All these segments are multiple of a predefined time unit, denoted as *time quantum* and derived from the local oscillator by applying a prescaler to a clock with rate *minimum time quantum*.

$$\text{time quantum} = m \times \text{minimum time quantum} \tag{1.1}$$

where m is the value of the prescaler. The *time quantum* is the minimum resolution in the definition of the bit time and the maximum assumed error for the bit-oriented synchronization protocol. The widths of the segments are defined by the standard as follows: SYNC_SEG is equal to 1 *time quantum*; PROP_SEG and PHASE SEG are between 1 and 8 times the *time quantum*; PHASE_SEG2 is the maximum between PHASE_SEG1 and the *information processing time*, which must be always less than or equal to twice the *time quantum*.

Two types of synchronization are defined: hard synchronization, and re-synchronization (Fig. 1.2).

- *Hard synchronization* takes place at the beginning of the frame, when the start of frame bit (see the frame definition in the following section) changes the state of the bus from recessive to dominant. Upon detection of this edge, the bit time is resynchronized in such a way that the end of the current quantum becomes the end of the synchronization segment SYNC_SEG. Therefore, the edge of the start bit lies within the synchronization segment of the restarted bit time.
- *Re-synchronization* takes place during transmission. The phase segments are shortened or lengthened so that the following bit starts within the SYNCH_SEG portion of the following bit time. In detail, PHASE_SEG1 may be lengthened or PHASE_SEG2 may be shortened. Damping is applied to the synchronization protocol. The amount of lengthening or shortening of the PHASE_SEG segments is upper bounded by a programmable parameter *resynchronization jump width*, which is set to be between 1 and min(4, PHASE_SEG1) times the *time quantum*.

Nodes can perform synchronization only when the bus state changes because of a bit value change. Therefore, the possibility of resynchronizing a bus unit during a frame transmission depends on the possibility of detecting at least one bit value transition in any interval of a given length. The bit stuffing protocol guarantees that the time interval between any two bit transitions is upper bounded (no more than 5 bits) in such a way that clocks should never drift beyond the possibility of recovery offered by the synchronization protocol.

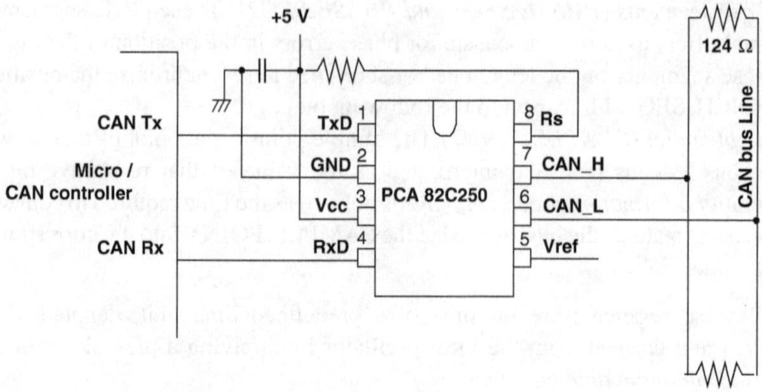

Fig. 1.3 The NXP PCA 82C250 transceiver and its connections

The device designer may program the bit-timing parameters in the CAN controller by means of appropriate registers. This selection is constrained by the size of the propagation delay segment depending on the maximum possible bus length at a specific data rate.

1.1.2 Transceivers, Network Topology and Bus Length

The interface between the CAN controller chip and the transmission medium (in most cases a two-wire differential bus) typically consists of a transmitting and a receiving amplifier: the CAN transceiver (*transceiver* stands for *transmit* and *receive*). The transceiver converts the electrical representation of the bit in use by the controller to that of the bus. As a transmitter, it provides sufficient output current to drive the bus electrical state, and protects the controller chip against overloading. As a receiver, it provides the recessive signal level and protects the controller chip input comparator against excessive voltages on the bus lines. Furthermore, it detects bus errors such as line breakage, short circuits and shorts to ground. Finally, it creates a galvanic isolation between a CAN node and the bus line.

Figure 1.3 shows an example of transceiver connections to the CAN controller and the CAN bus for the NXP PCA 82C250 transceiver.

Because of the definition of the bit time, a dependency between the bit time and the signal propagation delay exists, that is, between the maximum achievable bit rate (or transmission speed) and the length of the bus. The signal propagation delay to be considered for the computation of the maximum allowed bus length includes several stages, with variable delays, depending on the quality of the selected components: CAN controller (50–62 ns), optocoupler (40–140 ns), transceiver (120–250 ns), and cable (about 5 ns/m).

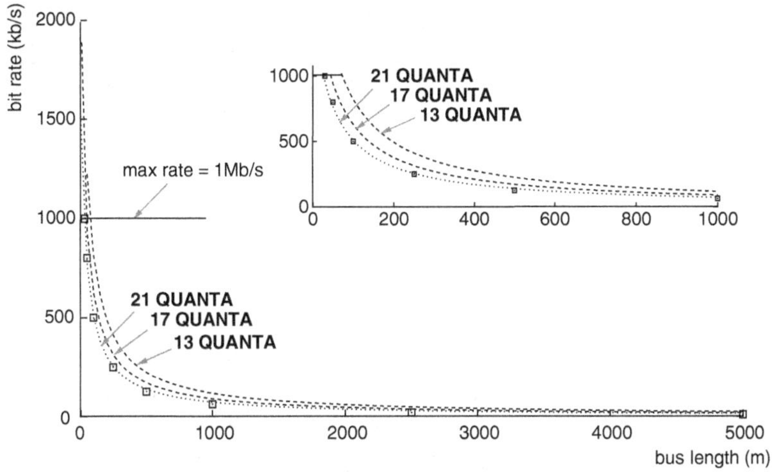

Fig. 1.4 Relationship between bus length and bit rate for some possible configurations

Table 1.1 Typical transmission speeds and corresponding bus lengths

Bit rate	Bit time (μs)	Bus length (m)
1 Mb/s	1	30
800 kb/s	1.25	50
500 kb/s	2	100
250 kb/s	4	250
125 kb/s	8	500
62.5 kb/s	16	1,000
20 kb/s	50	2,500
10 kb/s	100	5,000

For short (a few meters) high speed networks, the biggest limitation to bus length is the transceiver's propagation delay of the receiver. The relative importance of the electrical medium propagation delay increases when the bus length is increased. In many cases, however, the propagation delay is significant and the line resistance and the wire cross sections are key factors for achieving a high signal speed and the fastest possible bit rate when dimensioning a network. Figure 1.4 plots the correspondence between bus length and bit rate when the delays of the controller, the optocoupler, and the transceiver add to 250 ns, the cable delay is 5 ns/m and the bit time is divided in, respectively, 21, 17 or 13 times the *time quantum*, of which 8 (the maximum allowed) represents the propagation delay.

Some reference (not mandatory) value pairs for bus length and transmission speed are shown in Table 1.1, as well as in Fig. 1.4 as squared plots. The CANopen consortium has a similar table of correspondences (see Sect. 9.2).

In addition, the voltage drops over the length of the bus line as the length increases. The required cross section of the wire is calculated by the permissible voltage drop of the signal level between the two nodes farthest apart in the system

Table 1.2 Bus cable characteristics

Bus speed	Cable type	Cable resistance/m	Terminator	Bus length
50 kb/s at 1,000 m	0.75–0.8 mm² (AWG18)	70 m Ω	150–300 Ω	600–1,000 m
100 kb/s at 500 m	0.5–0.6 mm² (AWG20)	<60 m Ω	150–300 Ω	300–600 m
500 kb/s at 100 m	0.34–0.6 mm² (AWG22, AWG20)	<40 m Ω	127 Ω	40–300 m
1,000 kb/s at 40 m	0.25–0.34 mm² (AWG23, AWG22)	<26 mΩ	124 Ω	0–40 m

and the overall input resistance of all connected receivers. The permissible voltage drop must be such that the signal level can be reliably interpreted at any receiving node. The consideration of electromagnetic compatibility and choice of cables and connectors is also part of the system integrator's tasks. The system integrator is usually an automotive Original Equipment Manufacturer (OEM) connecting several sub-systems (ECU) using CAN as a backbone medium, or a sub-system integrator (a supplier) connecting fewer ECUs, or a wiring harness provider. Table 1.2 shows possible cable sections and types for selected network configurations.

1.1.2.1 Bus Termination

Electrical signals on the bus are reflected at the ends of the electrical line unless preventive actions are taken. Signal reflection occurs when a signal is transmitted along a transmission medium, such as a copper cable or an optical fiber. Some of the signal power may be reflected back to its origin rather than being carried all the way along the cable to the far end. This happens because imperfections in the cable cause impedance mismatches and non-linear changes in the cable characteristics. These abrupt changes cause some of the transmitted signal to be reflected. To avoid mismatches when a node reads the bus electrical status, signal reflections must be avoided. Terminating the bus line with a termination resistor at both ends of the bus and avoiding unnecessarily long stubs lines of the bus is the best corrective action. The largest possible simultaneous transmission rate and bus length line are achieved by using a structure as close as possible to single line structure and by terminating both ends of the line (this layout is also referred to as *linear*). Specific recommendations for this structure can be found in the related standards (i.e. ISO 11898-2 and -3). The method of terminating CAN hardware varies depending on the physical layer of the hardware itself (high-speed, low-speed, single-wire, or software-selectable). For high-speed CAN, both ends of the pair of signal wires (CAN_H and CAN_L) must be terminated. The termination resistors on the cable should match the nominal impedance of the cable.

ISO 11898 requires a cable with a nominal impedance of 120 ohms, and therefore 120 ohm resistors should be used for termination. If multiple devices are placed along the cable, only the devices on the ends of the cable need termination resistors.

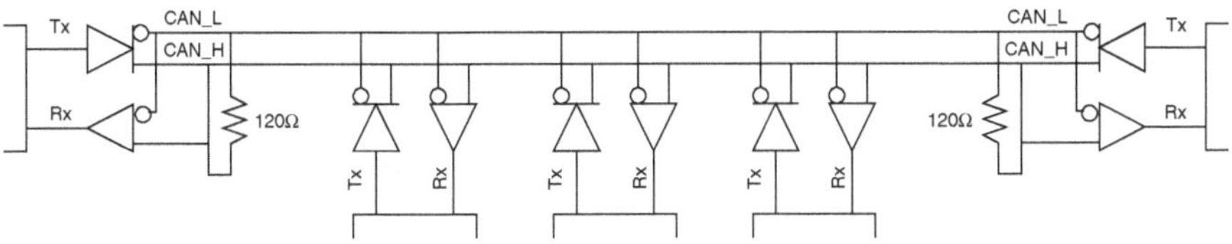

Fig. 1.5 Terminating a high-speed network

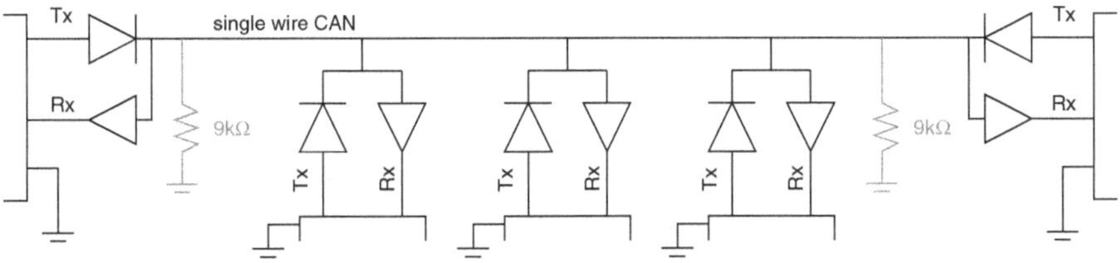

Fig. 1.6 Placement of termination resistors on a low-speed network

Figure 1.5 gives an example of how to terminate a high-speed network. The termination can be of basic type (a single resistor as in the figure) or of split type (two 60 ohm resistors with an intermediate capacitor between 10 nF and 100 nF connected to ground).

Unlike high-speed CAN, low-speed CAN requires termination on the transceiver rather than on the cable, meaning that each device on the network needs a termination resistor for each data line. Figure 1.6 shows where the termination resistors should be placed on a single-wire, low-speed CAN network. For example, the National Instruments single-wire CAN hardware includes a built-in 9.09 KΩ load resistor.

Using repeaters, bridges or gateways is a possible way to overcome the limitations of the basic line topology. A repeater transfers an electrical signal from one physical bus segment to another segment. The signal is only refreshed and the repeater can be regarded as a passive component comparable to a cable. The repeater divides a bus into two physically independent segments. This causes an additional signal propagation time. However, from the logical standpoint, it is one bus system. A bridge connects two logically separated networks on the data link layer (OSI layer 2). CAN identifiers are unique in each of the two bus systems. Bridges implement a store-and-forward mechanism for messages or parts thereof in an independent time-delayed transmission. Bridges differ from repeaters since they forward messages, which are not local, while repeaters forward all electrical signals including the CAN identifier. A gateway provides the connection of networks with different higher-layer protocols. Therefore, it performs the translation of protocol data between two communication systems. This translation takes place on the application layer (OSI layer 7).

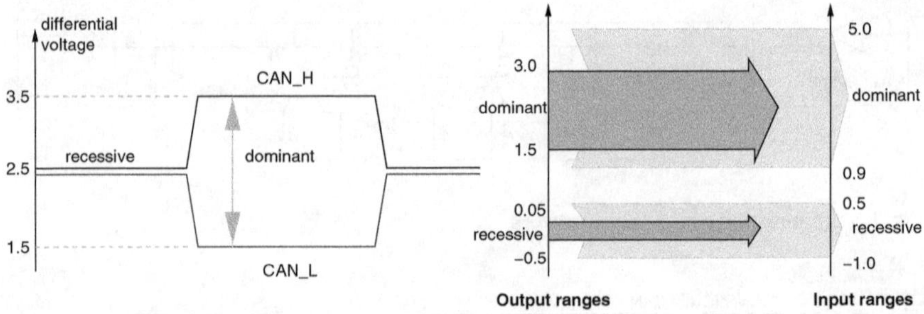

Fig. 1.7 Encoding of dominant and recessive bits

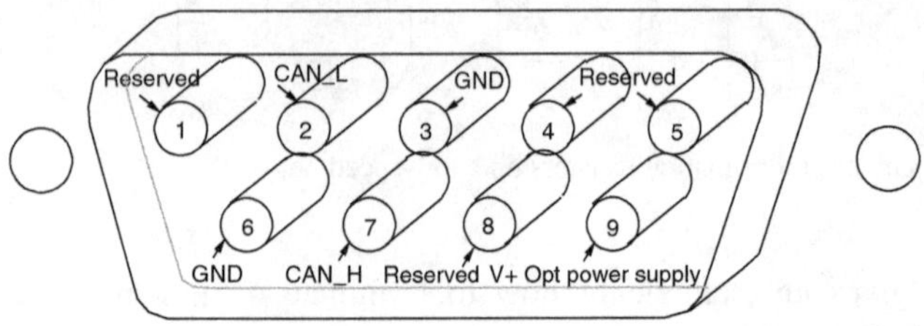

Fig. 1.8 The bus connector defined by CIA (DS 102-1)

1.1.3 Physical Encoding of Dominant and Recessive States

Controller Area Network specifies two logical states: recessive and dominant. According to ISO-11898-2, a differential voltage is used to represent recessive and dominant states (or bits), as shown in Fig. 1.7.

In the recessive state (usually logic 1), the differential voltage on CAN_H and CAN_L is less than the minimum threshold (<0.5V receiver input or <1.5V transmitter output). In the dominant state (logic 0), the differential voltage on CAN_H and CAN_L is greater than the minimum threshold. If at least one node outputs a dominant bit, the status of the bus changes to "dominant" regardless of other recessive bit outputs.

1.1.4 Connectors

Similar to the physical medium, connector types are also not defined by the standard but identified by other bodies. Figure 1.8 shows the pin layout of the very popular CIA DS 102-1 9-pin shielded connector.

1.1.5 The Physical Layer in ISO and SAE Standards

The implementations of the physical layer are based upon the following requirements:

- All nodes synchronize at the bit level whenever a transmission takes place.
- The arbitration mechanism (as defined in the following sections).
- All nodes agree on the logical value encoded by the electrical status of the bus (including the ability of representing "dominant" and "recessive" bits).

In principle, the system designer can choose any driver/receiver and transport medium as long as the Physical Signalling (PS) requirements of the protocol are met, including electrical and optical media, but also power lines and wireless transmission. In practice, the physical layer has been specified for specific class of users or applications by standardization bodies or (industrial) user groups.

ISO included CAN in its specifications as ISO 11898, with three parts: ISO 11898-1, ISO 11898-2 (high speed CAN) and ISO 11898-3 (low-speed or fault-tolerant CAN). ISO standards include the Physical Medium Attachment (PMA) and Medium Dependent Interface (MDI) parts of the physical layer (not defined by CAN). The most common type of physical signalling is the one defined by the CAN ISO 11898-2 standard, a two-wire balanced signaling scheme. ISO 11898-3 defines a fault-tolerant two-wire balanced signaling scheme for lower bus speeds, in which the signaling can continue if one bus wire is cut or shorted. In addition, SAE J2411 (SAE is the Society of Automotive Engineers) defines a single-wire (plus ground, of course) physical layer.

1.1.5.1 ISO 11898-2

ISO 11898-2 is the most used physical layer standard for CAN networks. The data rate is defined up to 1 Mbit/s with a required bus length of 40 m at 1 Mbit/s. The high-speed standard specifies a two-wire differential bus whereby the number of nodes is limited by the electrical busload. The two wires are identified as CAN_H and CAN_L. The characteristic line impedance is 120 Ω, the common mode voltage ranges from −2 V on CAN_L to +7 V on CAN_H. The nominal propagation delay of the two-wire bus line is specified at 5 ns/m. The SAE J2284 specification applies to automotive applications. For industrial and other non-automotive applications the system designer may use the CiA 102 recommendation. This specification defines the bit-timing for rates of 10 kbit/s to 1 Mbit/s. It also provides recommendations for bus lines and for connectors and pin assignment.

The ISO 11898-2 specification requires that a compliant transceiver meets a number of electrical specifications. Some of these specifications are intended to ensure the transceiver can overcome harsh electrical conditions, thereby protecting the communications of the CAN node. For example, the transceiver must be resilient

Table 1.3 Acceptable voltage ranges for CAN transmitters and receivers

Parameter	Min	Max	Unit
DC Voltage on CANH and CANL	−3	+32	V
Transient voltage on CANH and CANL	−150	+100	V
Common mode bus voltage	−2.0	+7.0	V
Recessive output bus voltage	+2.0	+3.0	V
Recessive differential output voltage	−500	+50	mV
Differential internal resistance	10	100	kΩ
Common mode input resistance	5.0	50	kΩ
Differential dominant output voltage	+1.5	+3.0	V
Dominant output voltage (CANH)	+2.75	+4.50	V
Dominant output voltage (CANL)	+0.50	+2.25	V
Output current		100	mA

to short circuits on the CAN bus inputs from −3 V to +32 V and to transient voltages from −150 V to +100 V. Table 1.3 shows the main ISO 11898-2 electrical requirements,

1.1.5.2 ISO 11898-3

This standard is mainly used in the automotive industry for body electronics. As the specification assumes a short network length, the problem of signal reflection is not as important as for longer bus lines. This makes the use of an open bus line possible. In ISO 11898-3 the bus topology is not limited to be strictly linear, but branching is allowed with a 180 Ω termination on each branch. Also, voltage levels do not need to be symmetric on both wires. As a limit case, it is possible to transmit over just one bus wire in case of an electrical failure of one of the lines. ISO 11898-3 defines data rates up to 125 kbit/s and maximum bus length proportional to the data rate used and the busload. Up to 32 nodes per network are specified. The common mode voltage ranges between –2 V and +7 V. The power supply is defined at 5 V. Fault-tolerant transceivers support error management including the detection of bus errors and automatic switching to asymmetrical signal transmission.

1.1.5.3 SAE J2411 Single Wire

The single-wire standard SAE J2411 can be used for CAN network applications with low bit rate and short bus length. The communication takes place via a single bus line at a nominal data rate of 33.3 kbit/s (83.3 kbit/s in high-speed mode for diagnostics). The standard defines up to 32 nodes per network. The main application area is in-vehicle body control networks. The bus medium is a unshielded single wire and branching is possible. The standard includes selective node sleep capability, which allows regular communication to take place among several nodes while others are left in a sleep state.

1.1.5.4 ISO 11992 Point-to-Point

The ISO 11992 standard describes an additional approach to using Controller Area Network low-speed networks with fault-tolerant functionality. It defines a point-to-point connection for use in applications such as in vehicles towing trailers, possibly extended to daisy-chain connections. The nominal data rate is 125 kbit/s with a maximum bus line length of 40 m. The standard defines the bus error management and the supply voltage (12 V or 24 V). An unshielded twisted pair of wires is defined as the bus medium.

1.1.5.5 Others

Optic fibers are among the physical signalling media that are not standardized. With optical media the recessive level is represented by "dark" and the dominant level by "light." Due to the directed coupling into the optical media, the transmitting and receiving lines must be provided separately. Also, each receiving line must be externally coupled with each transmitting line in order to ensure bit monitoring. A passive star coupler can implement this mechanism. The use of a passive star coupler is possible with a small number of nodes, thus this kind of network is limited in size. The extension of a CAN network with optical media is limited by the light power, the power attenuation along the line, and the star coupler, rather than by the signal propagation as in electrical lines. Advantages of the optical media are emission-free transmission and complete galvanic decoupling. The electrically neutral behavior is important for applications in electromagnetically disturbed environments.

1.2 Message Frame Formats

In CAN, there are four different types of frames, defined according to their content and function.

- *Data* frames, which contain data information from a source to (possibly) multiple receivers.
- *Remote* frames, which are used to request transmission of a corresponding (same identifier) Data frame.
- *Error* frames, which are transmitted whenever a node on the network detects an error.
- *Overload* frames, which are used for flow control to request an additional time delay before the transmission of a Data or Remote frame.

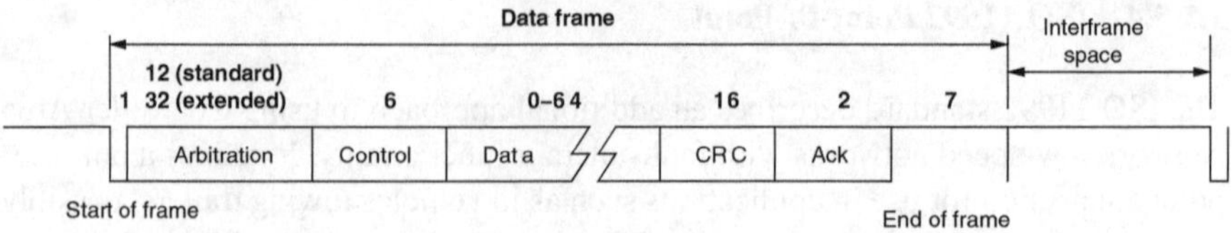

Fig. 1.9 The CAN data frame format

1.2.1 DATA Frame

Data frames are used to transmit information between a source node and one or more receivers. Data frames do not use explicit addressing to identify the message receivers. Instead, each receiver node states the messages that will be received based on their information content, which is encoded in the *Identifier* field of the frame. There are two different formats for CAN messages, according to the type of message identifier that is used by the protocol. Standard frames are frames defined with an 11-bit *Identifier* field. Extended frames have been made available from version 2.0 of the protocol as frames with an 29-bit *Identifier* field. Standard and extended frames can be transmitted on the same bus by different nodes or by the same node. The arbitration part of the protocol works regardless of the identifier version of the transmitted frames, allowing 29-bit identifier messages to be transmitted on the same network together with others with an 11-bit identifier.

The CAN data frame format is shown in Fig. 1.9, where the size of the fields is expressed in bits. Each frame starts with a single dominant bit, interrupting the recessive state of the idle bus. Then, the identifier field defines both the priority of the message for arbitration (Sect. 1.3) and the data content (identification) of the message stream. The other fields are: the *Control* field, containing information on the type of message; the *Data* field, containing the actual data to be transmitted, up to a maximum of 8 bytes; the *Checksum*, used to check the correctness of the message bits; the *Acknowledge* (ACK), used to acknowledge the reception; the *Ending delimiter* (ED), used to define the end of the message and the *Idle space* (IS) or *Interframe bits* (IF), used to separate a frame from the following one.

1.2.1.1 Identifier Field

The CAN protocol requires that all contending messages for the same medium have a unique identifier. The *Identifier* field consists of 11 (+1) bits in the standard format and 29 (+3) bits in extended format, following the scheme of Fig. 1.10. In both cases, the *Identifier* field starts with the 11 bits (the most significant bits, in the extended format) of the identifier, followed by the Remote Transmission Request (RTR) bit in the standard format and by the Substitute Remote Request (SRR) in the extended format. The RTR bit distinguishes data frames from remote request frames. It is dominant for data frames, recessive for remote frames. The SRR is

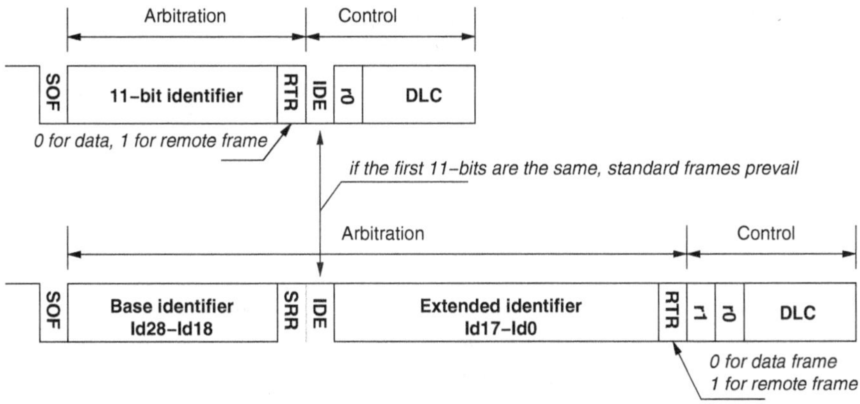

Fig. 1.10 The CAN identifier format

only a placeholder (always recessive) for guaranteeing the deterministic resolution of the arbitration between standard and extended frames.

The extended frame includes a single IDE bit (IDentifier Extension, always recessive), then the remaining 18 least significant identifier bits and finally, the RTR bit. The IDE bit is part of the control field in standard frames.

1.2.1.2 Control Field

The *Control* field contains 6 bits. The first two bits are reserved or predefined in content. In the standard message format the first bit is the IDE (Identifier Extension bit), followed by a reserved bit. The IDE bit is dominant in standard formats and recessive in extended formats. It ensures the deterministic resolution of the contention (in favor of standard frames) when the first eleven identifier bits of two messages (one standard, one extended) are the same. In the extended format there are two reserved bits. For these bits, the standard specifies that they are to be sent as recessive, but receivers will accept any value (dominant or recessive). The following four bits in the data frame define the length of the data content (Data Length Content, or DLC) in bytes. If the dominant bit is interpreted as 1 (contrary to the common notation in which it is read as 0) and the recessive as 0, the four DLC bits are the unsigned binary coding of the length (Fig. 1.11).

1.2.1.3 CRC + ACK Fields

The CRC and ACK fields are the next frame content. Their general layout is in Fig. 1.12. The CRC portion of the frame is obtained by selecting the input polynomial (to be divided) as the stream of bits from the start of frame (SOF) bit (included) to the *Data* field (if present) followed by 15 zeros. This polynomial is

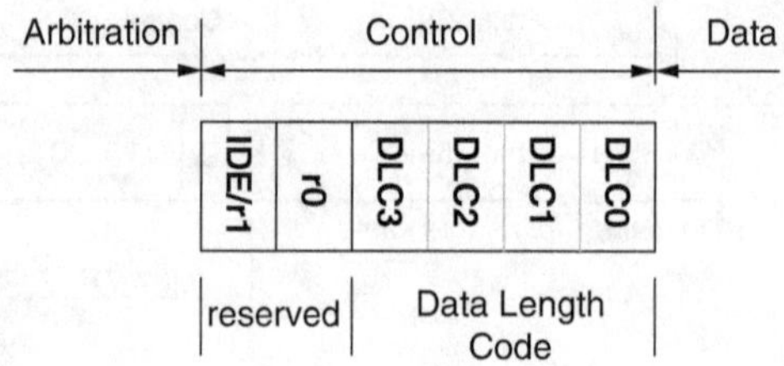

Fig. 1.11 The control field format

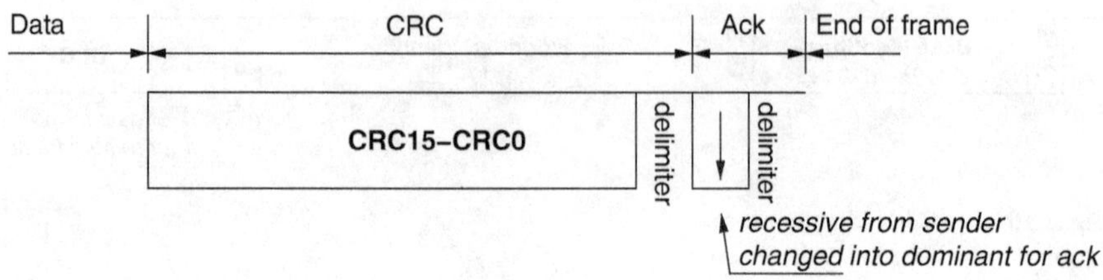

Fig. 1.12 The CRC and acknowledgement formats

divided by the generator

$$X^{15} + X^{14} + X^{10} + X^{8} + X^{7} + X^{4} + X^{3} + 1$$

The remainder of the division is the CRC sequence portion of the frame. Details on the computation of the CRC field (including the generating algorithm) can be found in the Bosch specification.

The ACK field consists of two bits. The first bit has the function of recording acknowledgments from receivers (ACK slot). The other bit is a delimiter (one bit of bus recessive state). The receivers that have validated the received message signal their acknowledgement to the sender by overwriting the recessive bit sent in the ACK slot by the transmitter with a dominant bit.

1.2.1.4 Interframe Space

Data frames and remote frames are separated by an *Interframe* space (7 recessive bits) on the bus.

The frame segments *Start of frame* (SOF), *Identifier*, *Control*, *Data* and *CRC* are subject to bit stuffing. The remaining bit fields of the data frame or remote frame (CRC delimiter, *ACK*, and *End Of Frame* (EOF) fields) are in fixed form and not stuffed.

1.2.2 Remote Frame

A remote frame is used to request the transmission of a message with a given identifier from a remote node. A remote frame has the same format of a data frame with the following characteristics:

- The *Identifier* field is used to indicate the identifier of the requested message.
- The *Data* field is always empty (0 bytes).
- The *DLC* field indicates the data length of the requested message (not the transmitted one).
- The *RTR* bit in the arbitration field is always set to be recessive.

1.2.3 Error Frame

The *Error* frame is not a real frame, but rather the result of error signaling and recovery action(s). The details of the error frame and its management are described in the following section on error management. The Error frame (not surprisingly given that it consists of 6 dominant or recessive bits in a sequence) is of fixed form with no bit stuffing.

1.2.4 Overload Frame

The purpose of the *Overload* frame is to inject a bus state at the end of a frame transmission that prevents the start of a new contention and then transmission. This allows an overloaded receiver to force a waiting period for the sender until it is once again ready to process the incoming data. The overload frame consists of 6 consecutive dominant bits. Its transmission starts during the first two bits of the *Interframe space* closing the transmission of the preceding frame. Since the interframe space should normally be of 3 bits, the overload signalling is detected by all other nodes, which also transmit overloads flags, possibly overlapping, up to a maximum of 12 dominant bits.

Newer controllers typically don't need to send overload frames, but the standard requires that they are capable of understanding and reacting to an overload frame. Of course, the *Overload* frame is also of fixed form with no stuffing.

1.3 Bus Arbitration

The CAN arbitration protocol is both priority-based and *non-preemptive*, as a message that is being transmitted can not be preempted by higher priority messages that are made available at the network adapters after the transmission has started.

The CAN bus essentially works as wired-AND channel connecting all nodes. The media access protocol works by alternating contention and transmission phases. The contention and transmission phases take place during the digital transmission of the frame bits.

At any time, if a node wishing to transmit finds the shared medium in an idle state, it waits for the end of the current bit (as defined by its internal oscillator), and then starts an arbitration phase by issuing a start-of-frame bit (a dominant bus state). At this point in time, each node with a message to be transmitted can start the competition to get access to the shared medium. All nodes will synchronize on the SOF bit edge and then, when the *Identifier* field starts, they will serially transmit the identifier (priority) bits of the message they want to send, one bit for each slot, starting from the most significant ones. Collisions among identifier bits are resolved by the logical AND semantics, and each node reads the bus state while it transmits.

If a node reads its identifier bits on the medium without any change, it realizes it is the winner of the contention and is granted access to transmit the rest of the message, while the other nodes switch to a listening mode. In fact, if one of the bits is changed when reading it back from the medium, this means there is a higher priority (dominant bit) contending the medium and thus the message is withdrawn by the node that has lost the arbitration. The node stops transmitting the message bits and switches to listening mode only.

Clearly, according to this contention resolution protocol, the node with the lowest identifier is always the winner of the arbitration round and is transmitted next (the transmission stage is actually not even a distinct stage, given that the message frame is transmitted sequentially with the fields following the *Identifier*). Given that in a CAN system, at any time, there can be no two messages with the same identifier (this property is also referred to as *unique purity*), the arbitration is deterministic and also *priority*-based, given that the message identifier gives in this case an indication of the message priority (the lowest identifier always wins).

An example of CAN bus arbitration is shown in Fig. 1.13. Three network nodes, labeled as Node 1 to Node 3 are trying to send the CAN messages with identifiers `0x15A`, `0x3D2` and `0x1F6`. The binary encoding is shown right below the hexadecimal message identifier values in the figure. The contention is started by the *Start Of Frame* bit, followed by the identifier bits. As the first identifier bit, all nodes send a 0 (dominant value). As a result, the bus state also shows the 0 value, all nodes see no difference with the transmitted value and proceed. At the next bit, nodes 1 and 3 send a bit at 0, while node 2 sends a bit at 1. The bus state stays dominant at 0. Node 2 realizes it tried to send a 1 and read a 0, it stops the transmission of the identifier bits of its message and switches to listening mode. The next bit sent by Node 1 and Node 3 has value 1 for both. The bus state will be 1 and both nodes continue. On the next bit, node 1 sends a zero and node 3 sends 1. Given that the resulting bus state is 0, node 3 withdraws from the contention and node 1 is the winner; it keeps transmitting until the end of the identifier and follows with the remaining fields of the frame.

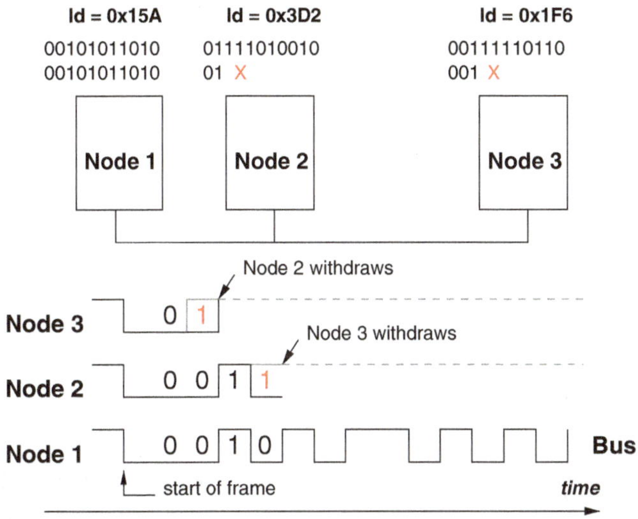

Fig. 1.13 An example of CAN bus arbitration based on message identifiers

1.4 Message Reception and Filtering

Controllers have one or more data registers (commonly defined as RxObjects) in which they store the content of the CAN messages received from the bus. Given that the protocol has no destination address field, the transmission semantics are the following. All transmissions are broadcasted, and nodes need a way to select the messages they are interested in receiving. This selection is based on the *Identifier* value of the frame, which not only defines the frame priority, but also gives information about the message content. Nodes can define one or more message *filters* (typically one filter associated with each RxObject) and one or more reception *masks* to declare the message identifiers they are interested in.

Masks can be individually associated to RxObjects, but most often they are associated to a group of them (or all of them). A reception mask specifies on which bits of the incoming message identifier the filters should operate to detect a possible match (Fig. 1.14). A bit at 1 in the mask register usually enables comparison between the bits of the filter and the received id in the corresponding positions. A bit at 0 means *don't care* or *don't match* with the filter data. In the example in the figure, the id transmitted on the bus is `01110101010` (`0x3AA`). Given the mask configuration, only the first, third, sixth, seventh and eight bit are going to be considered for comparison with the reception filters.

In the example in the figure, after comparing to the filters, the filter of RxObject1 is found to be a match for the required bits, and the incoming message is then stored in the corresponding RxObject.

The point in time when a message is considered to be valid is different for the transmitter and the receivers of the message. The transmitter checks all bits until the end of the *End of frame* field. Receivers consider a message as valid if there is

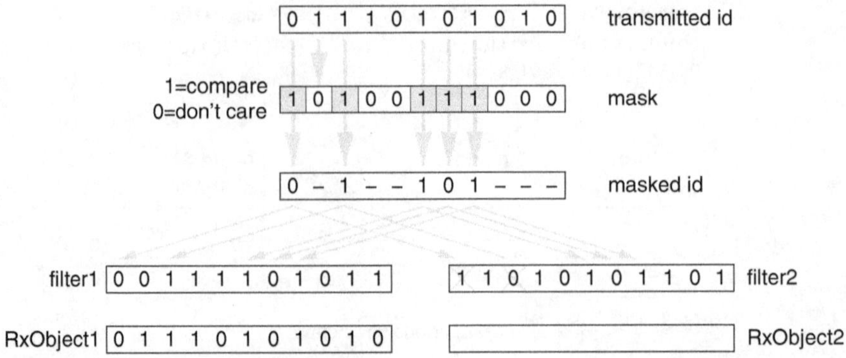

Fig. 1.14 Masks and filters for message reception

no error until the last but one bit of *End of frame*. The value of the last bit of END OF FRAME is treated as *don't care* (a dominant value does not lead to a FORM ERROR). The difference in the interpretation of a valid transmission/reception of a message can give rise to inconsistent message omissions/duplicates. This problem will be explored in detail in Chap. 6.

1.5 Error Management

The CAN protocol is designed for safe data transfers and provides mechanisms for error detection, signaling and self-diagnostics, including measures for fault confinement, which prevent faulty nodes from affecting the status of the network.

The general actions for error detection are based on the capability of each node of monitoring broadcast transmissions over the bus, whether it is a transmitting or a receiving node, and to report error conditions resulting from several sources. Corrupted messages are flagged by any node detecting an error. Such messages are aborted and will be retransmitted automatically.

1.5.1 CRC Checks

The CAN CRC has the following properties: it detects all errors with 5 or fewer modified bits, all burst errors (errors in consecutive bits) up to 15 bits long, and all errors affecting an odd number of bits. That specification states that multi-bit errors (6 or more disturbed bits or bursts longer than 15 bits) are undetected with a probability of 3×10^{-5}. Unfortunately, this evaluation does not take into account the effect of bit stuffing. It is shown in [61] that in reality, the protocol is much more vulnerable to bit errors, and even 2-bit errors can happen without being detected. More details will be provided in Chap. 6 on reliability.

1.5.2 Acknowledgement

As previously stated, the protocol requires that receivers acknowledge reception of the message by changing the content of the ACK field. In detail, upon successful reception, the first bit (ACK slot), which is sent as recessive, must be overwritten as dominant.

1.5.3 Error Types

The CAN protocol identifies and defines management for five different error types:

- *Bit* error. At any time, the node that has transmission rights, sends the message bits on the bus, but also monitors the bus state. A Bit error is detected when the bit value read from the bus is different from the bit value that is sent. Of course, the recessive bits sent as part of the arbitration process (in the *Identifier* field) or the ACK slot are not subject to the detection of bit errors.
- *Stuff* error. A Stuff error is detected at the sixth consecutive occurrence of the same bit in a message field that is subject to bit stuffing.
- *CRC* error. If the CRC computed by the receiver differs from the one stored in the message frame.
- *Form* error. A Form error occurs when a fixed-form bit field contains one or more illegal bits. As already stated, a receiver does not consider a dominant bit during the last bit of End of frame as a Form error. This is the possible cause of inconsistent message omission and duplicates, as further explained in Chap. 6 on reliability.
- *Acknowledgement* error. It is detected by a transmitter if a recessive bit is found on the ACK slot.

1.5.4 Error Signalling

The *Error* frame is not a true frame, but is actually the result of an error signaling sequence, consisting of the superposition of error flags, transmitted by different nodes, possibly at different times, and followed by an *Error Delimiter* field. Whenever a Bit error, a Stuff error, a Form error or an Acknowledgement error is detected by any station, the transmission of an *Error flag* is started at the respective station at the next bit boundary. Whenever a CRC error is detected, the transmission of an error flag starts at the bit following the ACK delimiter field, unless an error flag for another condition has already started.

At any point in time, based on their assumed reliability, nodes can be in *Error Active* or *Error Passive* state. When a node detects an error, if it is in *Error Active*

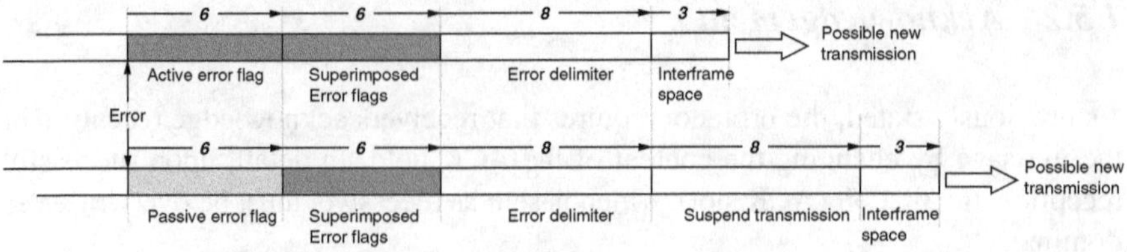

Fig. 1.15 Bit sequence after detection of an error

mode, sends an *Active Error* flag, consisting of six dominant bits. Otherwise, a *Passive Error* flag, consisting of six consecutive recessive bits is used for an *Error Passive* node.

The error flag form violates the bit stuffing rule and/or destroys the fixed form ACK or End of Frame fields. As a consequence, all other stations detect an error condition and start transmitting an error flag in turn. Hence, the sequence of dominant bits (which actually can be monitored on the bus) results from a superposition of different error flags transmitted by individual stations. The total length of this sequence varies between a minimum of six and a maximum of twelve bits. The passive error flag sent by *error passive* nodes has effect on the bus only if the error passive node is the transmitting node, which avoids a possible interference from the node on the transmissions of other nodes. Also, the error passive signaling node has to wait for six consecutive bits of equal polarity, beginning at the start of the passive error flag, before continuing. This additional delay clearly impacts its capability of participating in the very next contention, increasing the delays of its outgoing messages.

The recovery time from the time instant an error is detected, until the start of the next message is at most 31 bit times, that is, if there is no further error (Fig. 1.15).

1.5.5 Fault Confinement

The CAN protocol includes a fault confinement mechanism that detects faulty units and places them in passive or off states, so that they cannot affect the bus state with their outputs. The protocol assigns each node to one of three states:

- *Error active*: units in this state are assumed to function properly. Units can transmit on the bus and signal errors with an *Active Error* flag.
- *Error passive*: units in this state are suspected of faulty behavior (in transmission or reception). They can still transmit on the bus, but their error signaling capability is restricted to the transmission of a *Passive Error* flag.
- *Bus off*: units in this state are very likely corrupted and cannot have any influence on the bus. (e.g. their output drivers should be switched off.)

Units change their state according to the value of two integer counters, *Transmit Error Count* and *Receive Error Count*, which give a measure of the likelihood of a faulty behavior on transmission and reception, respectively. The state transitions are represented in Fig. 1.16.

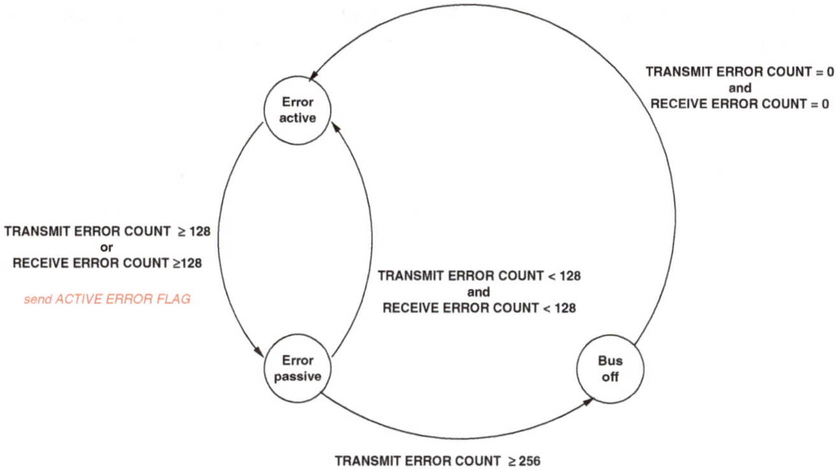

Fig. 1.16 Node error states

Table 1.4 The rules of updating transmit error count and receive error count

Node as receiver	Change	When node
Receive error count	+1	Detects an error unless a Bit error during an Active Error flag or an Overload flag
	+8	Detects a dominant bit as the first bit after sending an Error flag
	+8	Detects a Bit error while sending an Active Error flag or an Overload flag
	−1	Successful reception of a message if Receive Error Count ≤127,
	any n 119 ≤ n ≤ 127	Successful reception of a message if Receive Error Count >127
	+8	Node detects more than seven consecutive dominant bits after sending an Active Error flag, Passive Error flag or Overload flag (+8 for each additional 8 dominant bits)
Node as transmitter	**Change**	**When node**
Transmit error count	+8	Sends an Error flag (with some exceptions, see [16])
	+8	Detects a Bit Error while sending an Active Error flag or an Overload flag
	−1	Successful transmission of a message
	+8	Detects more than 7 consecutive dominant bits after sending an Active Error flag, Passive Error flag or Overload flag (+8 for each additional 8 dominant bits)

The transition from "Bus off" to "Active error" state is subject to the additional condition that 128 occurrence of 11 consecutive recessive bits have been monitored on the bus.

The counters are updated according to the rules of Table 1.4. A special condition can happen during the start-up or wake-up phases. If during start-up, a single node is on-line, and it transmits some message, it will get no acknowledgment, detect an error and repeat the message. Its state can become Error Passive, but will never go into the Bus Off state.

Chapter 2
Reference Architecture of a CAN-Based System

A CAN communication system requires the implementation of a complex protocol stack, from the application-level programming interface (API), down to the hardware implementation of the Medium Access Control (MAC) and Logical Link Control (LLC) layers inside the CAN adapter. Several properties of the communication, including latencies of messages, with possible priority inversions and loss of any type of time predictability, depend on the design choices at all levels including the hardware architecture of the peripheral adapter (e.g., available number of message buffers and their management), the structure of the operating system and middleware layers (e.g., the model of the operations for enqueuing and dequeuing messages in the device driver, OS, and Interaction layer), and the I/O management scheme (e.g., interrupt-based or polling-based) inside the device driver.

Figure 2.1 is a representation of the CAN communication architecture, as prescribed by the OSEK COM automotive standard [10] and appears, with little or no substantial change, in several commercial products. The main components of the architecture are:

- The *CAN Controller* is the (hardware) component that is responsible for the physical access to the transmission medium. It provides registers for the configuration of the connection to a bus with given characteristics, including the selection of the bit rate, the bit sample time, and the length of the interframe field. The controller also provides the functionality for managing all the aspects of the CAN protocol, including the management of the transmission modes and the handling of the bus off state. The controller offers a number of data registers for holding messages for outgoing transmissions and incoming data, and provides support for message masking and filtering on reception.
- The *CAN Device Driver* is the (software) component responsible for several tasks related to the low-level transmission of messages. The device driver initializes the CAN controller, performs the transmission and reception of individual messages, handles the bus-off state, and (when available) handles the sleep mode and the

M. Di Natale et al., *Understanding and Using the Controller Area Network Communication Protocol: Theory and Practice*, DOI 10.1007/978-1-4614-0314-2_2,

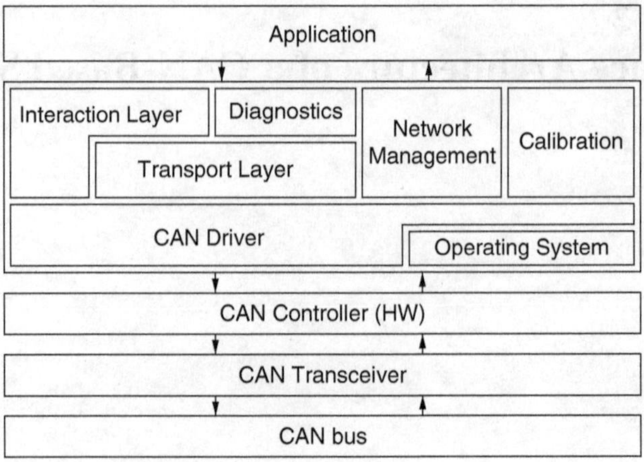

Fig. 2.1 Typical architecture of a CAN-based system. Actual systems may have different dependencies among layers

transitions in and out of it. Most CAN drivers also extend the controller capability with respect to the number of available transmission buffers, by providing software queue(s) for the temporary storage of messages.

- The *Transport* layer handles message segmentation and recombination whenever the upper layers, including the application-level functions, require the transmission of data with length exceeding 8 bytes. The transport layer provides mechanisms for sending acknowledgements and controlling the flow. The layer may also assist in establishing virtual channels and providing an extended addressing scheme.
- The *Interaction* layer offers an API to the application for reading and writing data according to application-defined types. In addition, it typically provides services for flow control, acknowledgement, and periodic transmission of messages. The interaction layer provides buffering and queuing of messages with several options. This layer also performs byte order conversion and message interpretation.
- The *Diagnostic and Network Management* layers determine and monitor the network configuration and provide information about the network nodes that are active. These layers also manage synchronized transition of all network nodes to the bus sleep mode when requested, and the management of virtual (sub)networks or network working modes.
- The *Calibration* layer supports the CAN Calibration Protocol (CCP). The CCP protocol was defined as a standard by the European ASAP automotive manufacturers working group (ASAP is an acronym from the German words for Working Group for the Standardization of Calibration Systems), and later handed over to the ASAM (Association for the Standardization of Automation and Measurement Systems). CCP is a high-speed interface based on CAN for measurement and calibration systems. The CCP protocol provides features for the development and testing phase as well as for end-of-line (flash) programming,

leveraging the general capability to read from and write into ECU memory locations for the purpose of parameter calibration and measurement.

The following sections provide additional details on the functionality, the design, the implementation options, and the management policies of the above hardware controllers and software layers.

2.1 CAN Controller and Bus Adapter

There are several CAN controllers available on the market. Some are available as stand-alone chips; however the most common option is that of a CAN controller integrated in a microcontroller unit, with a processing core, memory, and other devices. CAN controllers are of two types: *basic* and *full*.

In basic CAN controllers, hardware support is limited to the minimum logic necessary to generate and verify the bitstream according to the protocol. Only one transmission data register (or hardware buffer) is available for messages; if the node needs to transmit more than one message, then the upper layers need to buffer the messages in software queues. Two data registers are dedicated to the reception of messages. Hardware support for acceptance filtering is drastically limited in these controllers. Eight-bit code and mask registers are made available for the selection of the messages to be received. This means that filtering is limited to the 8 most significant bits (MSB) of the identifier. This might be barely sufficient in systems with 11-bit (standard) identifiers. When 29-bit identifiers are used, however, the hardware filtering must be supplemented by software. The two receiving buffers are usually accessed in FIFO (First-In-First-Out, the message that is extracted is the oldest among the two) order, so that a message can be received into one buffer while the microcontroller is reading the information from the other buffer. If the software does not respond in time to the reception interrupt signal and one more message is received while both receiving buffers are full, then the latest received message is overwritten and lost. Basic CAN controllers used to be popular because of their lower cost. Today, however, they are less common because the higher integration of electronic components allows the fabrication of full controllers at a very competitive price.

Full CAN controllers offer several *Objects* or data registers, typically configurable to act on the transmission or reception side. These controllers are capable of performing masking and filtering at the level of the individual reception registers (referred as *RxObjects*), and are also capable of arranging the transmission of messages in the (multiple) available transmission buffers (referred as *TxObjects*) according to a given policy. *RxObjects* and *TxObjects* are typically mapped in the RAM addressing space and made accessible to the controller DMA to free the processor from the burden of data transfer. Full CAN controllers are capable of performing full acceptance filtering for each reception object. In addition, if a remote (request) message is received, the controller is capable of performing an

Table 2.1 CAN controllers

Model	Type	Buffer Type	Priority and abort
Microchip MCP2515	Standalone controller	2 RX–3 TX	Lowest message ID, abort signal
ATMEL AT89C51CC03 AT90CAN32/64	8 bit MCU w. CAN controller	15 TX/RX msg. objects	Lowest message ID, abort signal
FUJITSU MB90385/90387 90V495	16 bit MCU w. CAN controller	8 TX/RX msg. objects	Lowest buffer num. abort signal
FUJITSU 90390	16 bit micro w. CAN controller	16 TX/RX msg. objects	Lowest buffer num. abort signal
Intel 87C196 (82527)	16 bit MCU w. CAN controller	14 TX/RX+1 RX msg. objects	Lowest buffer num. abort possible (?)
INFINEON XC161CJ/167 (82C900)	16 bit MCU w. CAN controller	32 TX/RX msg. objects (2 buses)	Lowest buffer num., abort signal
PHILIPS 8xC592 (SJA1000)	8 bit MCU w. CAN controller	One TX buffer	Abort signal

automatic comparison between the identifier indicated in the remote request and the identifiers of the messages in the *TxObjects*. If a match is found, the controller automatically sends the message. Of course, the software layers above the device driver are still responsible for ensuring that the data content of the message to be transmitted is up-to-date.

The configuration and management of the peripheral buffers is of utmost importance in the evaluation of the priority inversion at the adapter and of the (worst case) blocking times for real-time messages. For example, in [60], message latency is evaluated with respect to the availability of TxObjects for the messages, and the possibility of aborting a transmission request and changing the content of a TxObject.

A large number of CAN controllers is available in the market. Table 2.1 summarizes information on seven controllers from major chip manufacturers. The chips listed in the table are Micro Controller Units (MCUs) with integrated controllers or simple bus controllers. In case both options are available, the product codes of the integrated controller are shown between parenthesis. All controllers allow both polling-based and interrupt-based management of both the transmission and the reception of messages.

For the purpose of time predictability, controllers from Microchip [47] and ATMEL [14] exhibit the most desirable behavior (see the following chapter on worst-case timing analysis). These chips provide at least three transmission buffers, and the peripheral hardware selects the buffer containing the message with the lowest ID (the highest priority message) for attempting a transmission whenever multiple messages are ready. Furthermore, in those controllers, message preemption is always possible with the following behavior: the pending transmission communications are immediately aborted but the on-going communication will be terminated normally, setting the appropriate status flags.

Other chips, from Fujitsu [28], Intel [33], and Infineon [32], provide multiple message buffers, but the chip selects the lowest buffer index for transmission (not necessarily the lowest message ID) whenever multiple buffers are available. In [60] it is argued that, since the total number of available buffers for transmission and reception of messages is 14 (for the Intel 82527), the device driver can assign a buffer to each outgoing stream and preserve the ID order in the assignment of buffer numbers. Unfortunately, this is not always possible in several applications (including automotive) or when the node is acting as a gateway, because of the large number of outgoing streams. Furthermore, in several systems, message input may be polling-based rather than interrupt-based and a relatively large number of buffers must be reserved to input streams in order to avoid message loss due to overwrite.

Finally, the Philips SJA1000 chip [55], an evolution of the 82C200 controller discussed in [60], did not improve much on the shortcomings of its 82C200 predecessor, that is, a single output buffer, although with the possibility of aborting the transmission and performing preemption.

2.2 CAN Device Drivers

The CAN driver includes several functions that can be roughly classified into

- *Initialization and restart* for the initialization of the software data structures and the controller (hardware) registers both after power-on and after bus-off.
- *Transmission of messages*, providing for the transmission of a single message, and the handling of errors and exceptions (the Data Link Layer functionality).
- *Reception of messages*, both by interrupt or polling, with filtering and several queuing options on the receiving side.
- *Other functions*, including mode management, sleep states and wakeups.

The following sections provide details about the functionality and the data structures for each set of services.

2.2.1 Transmission

On the transmit side, the device driver software provides a function for requesting the transmission of a single CAN message. When the function is called, the driver first checks whether the transmitter is in the off-line state or not. Then, the driver checks the availability of a TxObject in the controller. If all the TxObjects are occupied by messages or not usable by the message to be transmitted (the allocation of TxObjects to messages is one of the policies under the control of the driver), the request can be stored temporarily in a software transmit queue. Some drivers may offer the option of evicting one of the messages from the adapter TxObjects to make it available for the newly arrived transmission request. Although this option makes sense if the transmission request is for a high-priority message (at the very minimum higher than one of the messages in the TxObjects), few drivers make it available. The number of available TxObjects depends on the specific CAN controller, but their assignment to messages is under the control of the CAN driver. TxObjects may be dedicated to messages in exclusive mode (avoiding the need to copy the DLC and ID fields) or associated with a set of messages to be put in a software queue. In any case, upon a request for transmission, if one TxObject is available, the message data is copied into it. Otherwise, the message is enqueued, and the transmit function typically returns with an acknowledgement code signalling that the transmit request was accepted by the driver. The message copy from the data buffer associated to the message to the CAN controller hardware register may be performed by a user notification function or by the CAN driver itself. If the transmission is rejected for any reason (error, absence of available TxObjects and/or no available software queue), then the application is notified and it is typically responsible for re-trying the transmit request.

After the message content has been copied into the adapter transmit buffer, transmission is enabled on the CAN controller. In most cases, completion of the transmission will be signalled by the arrival of an interrupt signal (if the driver enables the interrupt management of the empty transmit object event). A polling-based management is also possible; however this will be highly undesirable for time predictability. The interrupt service routine activated by the interrupt signal right after the successful transmission of a message performs a set of actions. First, the user has the option of setting up a confirmation mechanism, in the form of a confirmation flag (to be set by the driver upon reception), or a confirmation function (a notification function defined by the user to be called), or both. If they are used, then the confirmation flag is set and the confirmation function is called.

Next, if the CAN driver is configured to use a transmit queue, then it checks whether the queue is empty. If so, the transmit interrupt routine terminates. If there are messages waiting in the queue, one message is removed, copied into the available TxObject and the transmission of the message is enabled. The selection of the message to be extracted from the queue should be performed by priority (the message with the lowest identifier). Unfortunately, several drivers use a FIFO queue for the temporary storage of messages. The price paid for the apparent benefit of a simpler and faster management of the message queue is the possibility of multiple priority inversions and a very difficult or impossible time predictability, as explained in the next chapter.

In most drivers, the transmit queue holds only the transmission request of a CAN message but not the data content. The message data is stored in a dedicated buffer. On a transmit interrupt, when the driver extracts the message to be transmitted from the queue, it typically retrieves the DLC and Identifier (ID) fields of the message from the descriptor in the queue, and then copies the message data from its dedicated buffer area to the TxObject. The message copy is performed by the driver in the context of an interrupt service routine, but the same message buffer is also made available to the upper layers for updating the data content. It is typically the user's responsibility to guarantee the data consistency, because a write access by the upper layers (the application) may be interrupted by a transmit interrupt. The client software may ensure consistency by disabling interrupts while writing to the global buffer, or by using some other synchronization mechanism.

2.2.2 Reception

Controller Area Network messages are received asynchronously. For the upper software layers, message reception happens without any explicit service function call, either by using a notification function or asynchronously, through a mailbox. The CAN driver may perform the actual reception of the message (copying the contents from the RxObject buffer register and placing it in a suitable memory area) when it has been informed of the reception by the CAN controller. The controller can either send an interrupt signal, or the driver can be programmed to periodically check the content of the RxObjects using a polling strategy. In the rest of the section, we will assume an interrupt-based management of message reception. The polling option is available in several driver architectures, and the respective effects on the timing performance of messages will be discussed in later chapters.

When the receive interrupt signal arrives and the handler routine is executed, a CAN message has passed the masking and filtering acceptance scheme defined in the adapter hardware, and its contents are stored in a receive register. Unfortunately, in the case of a basic CAN controller, this is not sufficient to guarantee that the message is actually meant to be processed by the node. Also, in all controllers, there is the possibility that the node is acting as a gateway for the message, and needs to process it for forwarding in the shortest possible time. To this purpose, many drivers offer the opportunity of defining a receiver callback routine to be called after the hardware acceptance stage. If the user defines such a function, it may perform application-specific filtering and message processing. In addition, the driver typically offers a default service for software acceptance filtering using linear search, hash-table search or index search. If a CAN message has passed the hardware filter but is rejected by the software acceptance functions (in the case of a *basic CAN* receiving object), a special callback function may be called (if configured). If the received CAN identifier matches the identifiers in the table, the driver continues with the processing of the message received by copying it into ring buffers, FIFOs, and/or a global data buffer. Several CAN drivers optimize the memory required by copying only those bytes used by the application (up to the last byte within the message which contains data of interest for the application). In addition, indication actions may be defined for each received message. If this is the case, the indication flag is set and/or the indication function is called. The application has the responsibility of clearing the indication flag.

Regardless of the number of objects (transmit or receive registers) that are available on the CAN chip, in several current architectures (specifically in the automotive domain) message input is typically polling-based rather than interrupt-based, and all peripheral objects or buffers, except one, are reserved to input streams in order to avoid message loss by overwriting. In these systems, a single transmit object is typically available for all the output messages from a node.

2.2.3 *Bus-Off and Sleep Modes*

The CAN driver also has the responsibility to notify a detected bus-off state to the application by calling a special callback function. When a node goes in bus-off state, the application may attempt to restore its functionality by re-initializing the CAN controller using a suitable driver function.

Some CAN controller supports a sleep mode with reduced power consumption. In these cases, the driver provides service functions to enter and leave the sleep mode on the request of the application software. If the CAN controller can be automatically awakened by the CAN bus, then a callback is typically made available to the application to inform it of the node wakeup and to make the necessary arrangements.

2.3 Interaction Layer

The description of the typical Interaction Layer (IL) services and structure given in this section follows to a large extent the specification provided for this layer in the OSEK COM standard [10]. On the transmission side, the interaction layer provides:

- signal-oriented and application data-oriented functions to send data over a bus to a given application-level object destination;
- several transmission modes to the application for sending periodic streams or performing transmission on request (with the possibility of specifying a minimum time interval between any two transmissions);
- a notification to the application when a signal is transmitted, which acts as confirmation.

On the reception side, the layer

- provides the application with functions for the reception of application-level data arriving over the CAN bus;
- notifies the application when a new value for a given signal arrives as part of a message payload (indication);

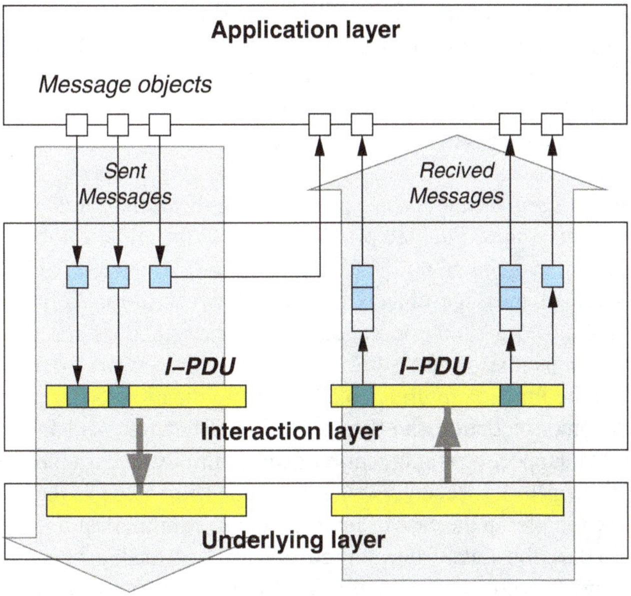

Fig. 2.2 Interaction layer architecture

- allows to control and monitor the reception of periodic data streams (or time-sensitive signals) by allowing the definition of timeouts on reception and callbacks on timeout expiration;
- provides (in case of an error) notification to the application and allows the definition of default values to be set in the reception mailboxes.

The IL also provides support for multiple channels in transmission and reception, and multiplexing (demultiplexing) of information. In OSEK COM, the IL realizes the abstraction of communication from the physical placement of senders of receivers, offering the same API for local as well as remote communication.

Administration of messages is done in the IL based on the concept of *message objects* (in the OSEK COM terminology, also defined as *application signals* in some commercial products). Message objects are associated with application types on both the sending and receiving sides, and have an associated identifier. OSEK COM supports only static object addressing. A statically addressed message object can have multiple receivers (or even no receiver at all!) defined at system generation time. A message object can have a static length or its length may be variable with a maximum defined at design-time (in this case, it is a *dynamic-length* message object). The IL packs one or more message objects into assigned *Interaction Layer Protocol Data Units* (I-PDU), and passes them to the underlying layer (Fig. 2.2). Within an I-PDU, message objects are bit-aligned. The size of an object is specified in bits, and the object is allocated to contiguous bits within its I-PDU. Objects cannot be split across I-PDUs.

The byte order (endianess) in a CPU can differ from other CPUs on the network. Hence, to ensure correct interpretation of message data and to provide interoperability across the network, the IL defines a network representation and provides for the conversion from the network representation to the local CPU representations and vice versa. The conversion is typically statically configured on a per-message basis.

The IL offers a set of functions as part of an application programming interface (API) to handle messages. The API provides services for initialization, data transfer, communication management, notification, and error management. The functions for the transmission of message objects over the network are usually non-blocking. This implies that a function for message transmission does not return a value indicating the transmission status/result because the network transfer is still in progress when the function returns. To allow the application to determine the status of a transmission or reception, the IL provides a notification mechanism.

OSEK COM supports communication from multiple sender objects to multiple receiver objects (m-to-n). Receivers may be on the same node as the source object, or they may be remote. In the case of internal (local) communication, the Interaction Layer (IL) makes the data from source object(s) immediately available to the destination objects (zero, one, or multiple) by direct copy. In the case of remote communication, the data from sender object(s) are copied into zero or one I-PDUs.

I-PDUs may contain data from one or more sending objects and can be received by zero or more CPUs. A receiving CPU reads the I-PDU data content and forwards the data to the destination (zero or more) receiving objects, where the data values become available to the application software. A receiving message object may receive the message from a local sending object or an I-PDU (in case it receives data from the network).

A receiving message object can be defined as either *queued* or *unqueued*. In the case of queued objects, the queue is of FIFO type. The queue size needs to be specified individually for each object. When the queue is empty, the IL does not provide any message data to the application. An object can only be read once, i.e., the read operation removes the oldest object from the queue. If the queue of a receiving message object is full and new object data arrive, the new data is lost and the next request for reception on the IL object returns the (oldest) data values and also the information that a message has been lost. The unqueued message objects can be read multiple times and are not consumed. An arbitrary number of receivers may receive the message object and each of them will get the last received value. This means that unqueued objects are overwritten by newly arrived messages. If no message has been received since the system startup time (the initialization of the IL layer), then the application receives the message value set at initialization. Given that the object values are shared, the IL needs to guarantee that the data in the application's message copies are consistent using a suitable locking or synchronization mechanism.

On the transmission side, an object to be transmitted on the network can have one of the following two transfer properties:

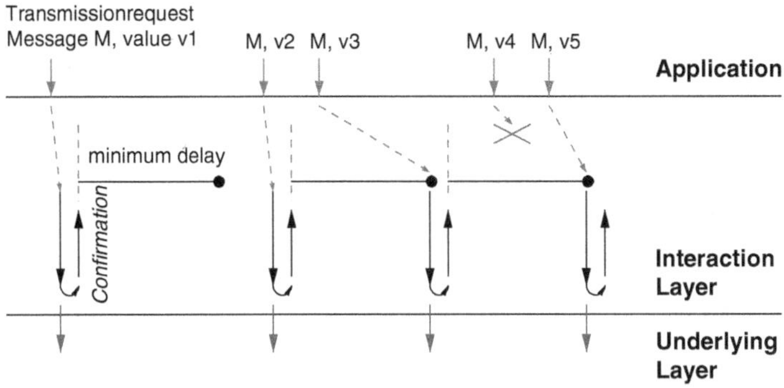

Fig. 2.3 Timing for the transmission of direct mode message objects

- *Triggered Transfer*: the data field in the assigned I-PDU is updated with the object contents, and the IL issues a request for the I-PDU transmission.
- *Pending Transfer*: the object value in the I-PDU is updated, but the I-PDU is not transmitted immediately.

These reflect into three transmission modes for I-PDUs, which are:

- *Direct Transmission* mode: the transmission of the I-PDU is initiated by the request for the transmission of a triggered transfer object.
- *Periodic Transmission* mode: the I-PDU is transmitted periodically by the IL without the need of an explicit request (for pending objects).
- *Mixed Transmission* mode: the I-PDU is transmitted using a combination of both the direct and the periodic transmission modes.

2.3.1 Direct Transmission Mode

The transmission of an I-PDU with direct transmission mode originates from the request for transmission of any object with triggered transfer property mapped into the I-PDU. The request immediately results in a transmission request for the I-PDU from the IL to the underlying layer. Each I-PDU with a direct transmission mode is also associated with a *minimum delay time* between transmissions. The minimum delay time starts at the time the transmission is confirmed from the lower layers. Any request for transmission arriving before the expiry of the minimum delay time is postponed and only considered at the end of the delay interval. As shown in the Fig. 2.3, the value `v1` is assigned to message object `M`, which is transmitted in direct mode. The message is copied in an I-PDU, which is immediately transmitted using the lower level API. Upon reception of the confirmation, the minimum delay timer is started. The following request for transmission for the same object with value `v2` comes after the minimum delay, and is therefore immediately processed.

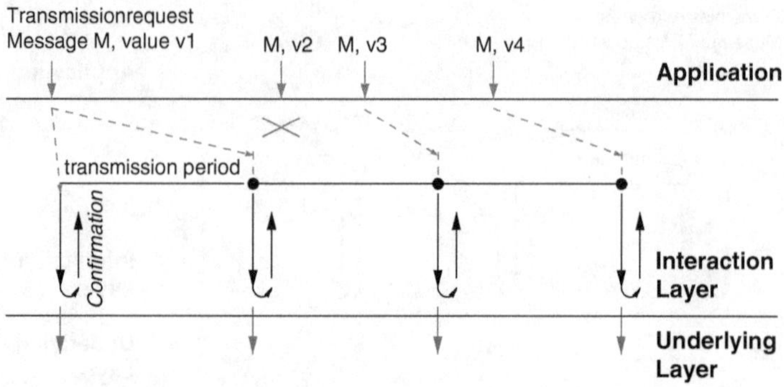

Fig. 2.4 Timing for the transmission of periodic mode messages objects

This is not true for the request associated with value `v3`. In this case, the minimum delay has not expired yet, and the I-PDU transmission will wait until the timer expires. Similarly, the request for `v4` comes before the minimum delay has passed from the previous transmission. In this case, however, before the transmission of the I-PDU containing the value `v4` can be issued, a new value `v5` is written into the object, and the value `v4` is overwritten and lost. The minimum delay time does not apply to the case of transmissions resulting in an error, e.g., expiration of the transmission deadline. In this case, a retransmission attempt can start immediately.

2.3.2 *Periodic Transmission Mode*

In the periodic transmission mode, the IL issues periodic transmission requests for an I-PDU to the underlying layer. On the application side, the API calls for the transmission of message objects (`SendMessage` and `SendDynamicMessage` in OSEK COM) to update the message object fields of the I-PDU with the latest values to be transmitted, but do not generate any transmission request to the underlying layer. The periodic transmission mode ignores the transfer property of the objects contained in the I-PDU, although clearly only pending transfer objects should be associated with a periodic transmission mode. The transmission is performed by repeatedly calling the appropriate service in the underlying layer with a period equal to the one defined for the periodic transmission mode. Figure 2.4 shows an example of periodic transmission modes. Regardless of the time at which the contents of the sending objects are updated, the I-PDU is transmitted periodically to the lower layers. This implies that some values may be transmitted twice (as is the case for `v1` in the figure), or may be overwritten (as happens to `v2`) and never transmitted over the network.

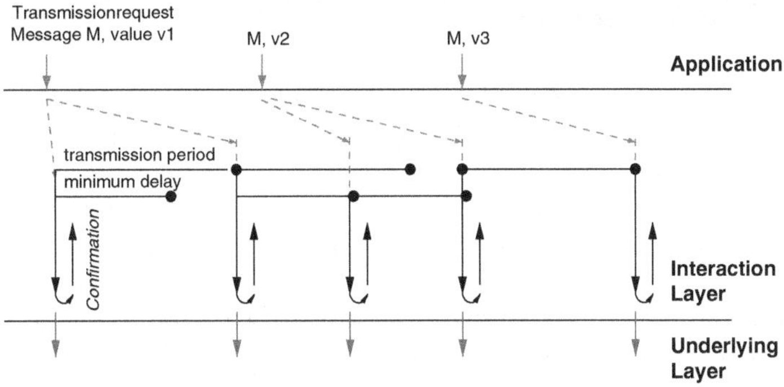

Fig. 2.5 Timing for the transmission of mixed mode messages objects

2.3.3 *Mixed Transmission Mode*

The mixed transmission mode is a combination of the direct and the periodic modes. Similar to the periodic mode, transmissions are performed with a given period by calling the appropriate service in the underlying layer. However, I-PDUs may also be transmitted upon request in between the periodic transmissions. These requests occur on-demand with the same modality of the direct transmission mode, that is, whenever the application requests the transmission of an object with triggered transfer property mapped into the I-PDU. Intermediate transmissions do not modify the base cycle, but must still comply with the minimum delay time defined between any two transmissions. If a transmission is requested within a shorter time, the I-PDU transmission is postponed until the delay time expires. The minimum delay interval also applies to periodic transmissions. Therefore, a periodic transmission request that occurs before the minimum delay time from any direct transmission will be delayed to ensure that no two transmissions are separated by less than the minimum delay. The following periodic transmission instances will occur after a full period has elapsed from the delayed instance (the entire periodic sequence is shifted in time). An example is shown in Fig. 2.5. The value `v1` is transmitted twice with the regular period of the periodic mode. Then, the value `v2` is transmitted, but not immediately. The I-PDU transmission is postponed to allow a minimum delay from the last periodic instance. Also, the next periodic transmission (again with value `v2`) is deferred to allow for a minimum delay with respect to the last direct transmission.

The periodic transmission sequences of the periodic and the mixed transmission modes are activated by a call to a suitable API service. This call also typically allows to specify the offset for the transmission requests (the time interval to the first transmission instance). Correspondingly, the periodic transmission mechanism is stopped by means of another API function.

2.3.4 Deadline Monitoring

OSEK COM provides a mechanism for monitoring the transmission and reception times of message objects and taking appropriate actions in the case of timing faults. This mechanism is called *Deadline Monitoring*. On the sender side, deadline monitoring verifies that the underlying layer confirms the request for a transmission within a given time period. On the receiver side, the IL checks that periodic messages are received within the time period associated with them. Deadline monitoring applies to external communication only and the monitoring is performed at the level of the I-PDU, that is, controlling the reception of the I-PDU containing the object. Deadlines and actions can be configured individually for each object. The use of this mechanism is not restricted to monitoring the reception of messages (I-PDUs) transmitted using the periodic transmission mode, but also applies to I-PDUs sent using the direct and the mixed modes.

2.3.5 Message Filtering

The IL also provides filtering functions that extend and complement the base filtering mechanisms offered by the adapter and driver layers, based on the identifier of the received message. IL filtering is performed for each object on both the sending and the receiving side. Filtering provides means to discard the message objects received or to be sent when a set of conditions are not met by the message object based on its value content or other attributes. Each message filter is a configurable function constructed by composing a set of predefined algorithms. On the sender side, the filter algorithm decides whether to actually update the I-PDU and send it according to the message contents (for example, avoiding the transmission if the value of the object has not changed). There is no filtering on the sender side for internal transmissions. On the receiver side, a filter mechanism may be used for both internal and external transmission. Filtering is only used for fixed-size messages that can be interpreted as C language unsigned integer types (characters, unsigned integers, and enumerations). In OSEK COM, the filter algorithm can use a set of parameters and values, including the current and last values of the message object, a bitmap used as a mask, a minimum and maximum value, the object period and offset, and the number of occurrences of the message.

2.4 Implementation Issues

In order to complete the discussion about the typical structure and features of a CAN architecture, it is useful to discuss some of the implementation options in more details. As specified in the previous sections, the CAN driver and interaction layers contain a number of options, protocol specifications (including management of time for timeouts and periodic transmission) and data structures (for queues and buffer storage) for which design, algorithm and code solutions may differ, and the implementation choices may impact the timing performance and the predictability of the message transmission times.

2.4.1 Driver Layer

On the transmission side, the driver puts the messages that need to be sent in a priority based software queue. The queue, sorted by message identifier (priority), is used to assign the transmit object inside the adapter. As described in the previous section, the driver can be configured to provide access to the transmission objects in exclusive mode, or it may associate a software queue to them. In most cases, the output of a software queue is not associated to a set of TxObjects but only to one, with the possibility of setting up multiple queues, one for each TxObject.

As for the implementation, the message queue should be a priority queue, in such a way that whenever a TxObject becomes available, the highest priority (lowest identifier) message is extracted from the queue and copied into it. FIFO queues can easily result in large priority inversions and undesirable message delays in the worst case. For the implementation of a priority queue, different options exist. The simplest and usually inefficient way is a simple sorted list. If the list is kept sorted, the time to add an element (message) to the list (insertion time) is in the order of $O(n_m)$ (the insertion time depends on the number n_m of messages in the queue) while extraction is done in $O(1)$ (constant) time, given that the highest priority message is always the first in the queue. To get better performance, priority queues may be realized as heaps, giving $O(\log(n_m))$ performance for insertions and extractions of messages. There are several heap structures and management algorithms available, including self-balancing binary search trees, binomial heaps, and Fibonacci heaps. The heap structure can be further optimized based on the knowledge of the set of messages that are transmitted by the node. To this purpose, it is worth mentioning that in most cases, the configuration of the CAN driver data structures or even procedures is performed automatically by code generation and configuration tools. Given that in the case of a CAN message queue the values to be sorted in the priority list are known, faster implementations are possible. In general, time efficiency can be traded for memory space. If the largest priority value to be managed is P (integer), a van Emde Boas tree [54] can be used to perform insertion and extraction operations in $O(\log(\log(P)))$ time, at an additional memory cost in the order of $O(\log_2(P))$. Constant time insertions and extractions are also possible at a memory cost $O(P)$. Both methods use bitmaps to store information about the priority values that are currently enqueued.

Other driver implementation issues that are worth discussing are the difference between implementations supporting the *interrupt* based (which is the most common method) and the *polling* based management of transmissions. Messages are

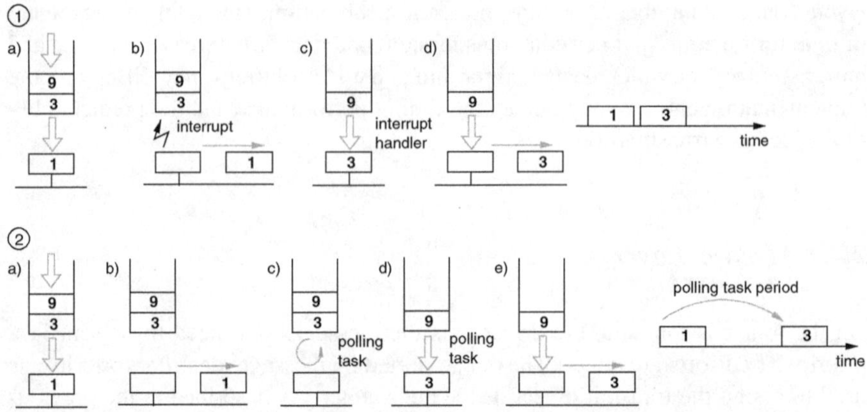

Fig. 2.6 Interrupt- or polling-based management of the TxObject at the CAN peripheral

dequeued by an interrupt handler, triggered by the interrupt signal that informs the completion of the transmission of the message occupying the TxObject at the adapter (shown in sequence 1 of Fig. 2.6), or by a polling task (shown in sequence 2 in Fig. 2.6).

- *Interrupt based*: The interaction layer (or the application if there is no interaction layer) use the driver API function to send messages. Inside the driver, if a TxObject that is a valid destination for the message is empty, it is filled with the message data, otherwise the message is queued. An example is shown in step 1a of the figure, where messages with identifiers 1, 3, and 9 are queued. When a message wins the contention (possibly after a *contention delay*) and leaves the TxObject, the buffer becomes empty and an interrupt is triggered (step 1b). Messages are dequeued by the interrupt handler, executed in response to the interrupt signal. The interrupt handler selects the message on top of the queue (the highest priority message) as the one that has to be placed into the buffer (step 1c). After the transmission of this message, the situation on the bus is the one on the right hand side of the sequence (1): the transmission of the messages that are placed in the queue is typically almost back-to-back, with a possible small separation time that is caused by the response time of the interrupt handler.
- *Polling based*: Messages are queued by the IL as in the previous case. In addition, an I/O driver-level task is activated periodically. When a message is transmitted, the next message in the queue is not immediately fetched to replace it. Only when the period of the polling task expires, it checks if the peripheral buffer is empty (step 2c of the second sequence). If so, it fetches the highest priority message from the queue and transfers it into the available buffer (step 2d). In this case, the typical behavior is that the transmission of messages from the same node is separated by a time interval roughly equal to the period of the polling task (possibly with network idle time in between).

On the receive side, the reception of a message causes a call of the Rx-Interrupt. The interrupt sub-routine compares the incoming message with the existing one. Depending on the result of the comparison, a special flag associated with the message and indicating the status of Changed Data may be set (to enable conditional application-level processing when new values are available). The interrupt sub-routine will copy the message to the buffer of the Interaction Layer.

2.4.2 *Interaction Layer*

Most of the functionalities required from the interaction layer requires the execution of periodic or time-driven computations. Some of the functionalities check the requests for data transmission, realize the periodic communication for the periodic mode I-PDUs and enforce the minimum delay for all messages. In several implementations these functions are realized by code called periodically at a fixed time interval and executed as part of a task in a multitasking operating system. We call this task *TxTask*, in accordance with the name it is assigned in several commercial solutions [4]. In most cases, the TxTask not only processes and performs the periodic transmissions of I-PDUs by calling the lower layer transmission API functions, but also synchronizes the transmission of all messages. In this case, transmissions on request will just store a transmit request. The actual transmission will be processed by the TxTask and therefore deferred until the next call of TxTask. If the TxTask period equals the minimum delay time, this ensures compliance with the IL protocol requirements for a minimum inter-transmission time. Other tasks to be performed periodically include timeout monitoring. These tasks are performed by a different function, typically called with a fixed period and implemented as part of the TxTask (if the periods are the same) or another task.

Application tasks copy the data values for all the signals that need to be transmitted in variables shared with the TxTask (see Fig. 2.7). Inside the IL, the TxTask is activated periodically and, at some point (typically at the end) of its execution, calls the driver- (or transport-) level function for the transmission of the I-PDU. In many cases, the I-PDUs are limited to be less than 8 bytes, therefore avoiding the need for a transport layer implementation and mapping one-to-one to CAN messages at the driver level.

In addition, each message M_i is defined to be transmitted only in some *virtual network* configurations, corresponding to operating modes of the CAN bus, and has an associated period T_i, expressed as an integer multiple of the TxTask period $T_{\mathrm{TX}}(T_i = k_i T_{\mathrm{TX}})$. Virtual networks may be defined for the system as the specification of a working mode, in which a subset of the system messages are actually transmitted, with their rates. The set of messages to be transmitted and their periods may be different from one virtual network configuration to another. Virtual networks management is typically part of the network management layer.

Inside the Interaction Layer, a descriptor consisting of a binary moding flag and a counter is associated to each message. The flag indicates whether the message

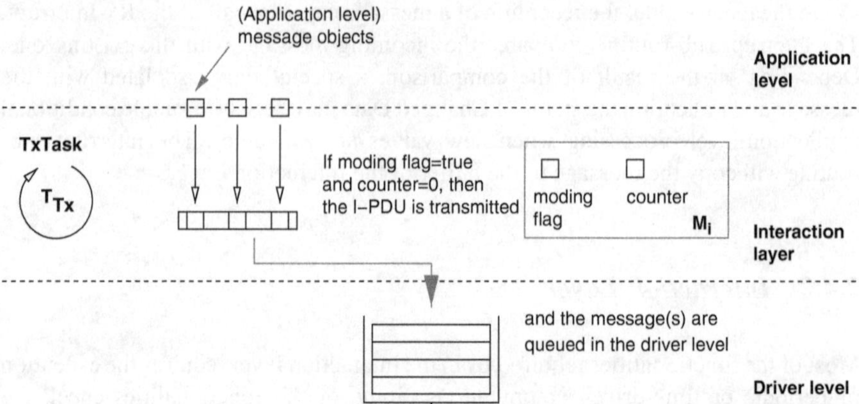

Fig. 2.7 The middleware task TxTask executes with period T_{TX}, reads signal variables are enqueues messages

should be sent for the currently active mode or not. When the TxTask is activated, it scans all the local message descriptors. For each of them, if the moding flag is set, it decrements the counter. If the counter is zero, it reads the values of all the signals that are mapped into the message, assembles the message data packet, and enqueues the packet in the software priority queue.

Some implementations also enforce a periodic processing for the received messages. On the receive side, the interaction layer buffers are used by the driver (often the interrupt handler inside it) to copy the content of the messages that passed the receive filtering at the driver level. The interrupt routine may also signal to the interaction layer the occurrence of a new value as compared to the latest reception (using a flag), and use an indication flag for signalling the occurrence of a newly received message. Inside the IL, a periodic task (called *RxTask*, clearly because of its association with the reception process) will check the occurrence of indication flags on the IL buffers. If the flag is set, the task resets the timeout timer for the message objects and calls the object-related indication function at the application level.

Chapter 3
Worst-Case Time Analysis of CAN Messages

Designers of CAN-based systems are of course interested in being able to predict the time performance of the messages exchanged over the network. The CAN protocol adopts a collision detection and resolution scheme, where the message to be transmitted is chosen according to its identifier. When multiple nodes need to transmit over the bus, the lowest identifier message is selected for transmission. The CAN MAC arbitration protocol encodes the message priority into the identifier field and implements priority-based real-time scheduling of periodic and aperiodic messages. Predictable scheduling of real-time messages on the CAN bus is then made possible by adapting existing real-time scheduling algorithms to the MAC arbitration protocol or by superimposing a higher-level scheduler designed purposely.

The deterministic and priority-based contention resolution should imply that the time performance of a message can be analyzed, at least to some degree. In most CAN systems nodes are not synchronized and the exact latency of a message depends on several factors, including the possibility that the message is delayed on the local queue or on the network adapter TxObjects because of interference from higher priority local or remote messages. The possibility of remote interference, with the corresponding delay, can be analyzed by identifying the worst-case scenario or by applying stochastic or statistical analysis.

Indeed, there are several ways of looking at the time performance of a message (or a system of messages). For safety-critical systems, or in those knowledge of the *worst-case latency* (sometimes also referred to as *worst-case response-time* for analogies with task scheduling on processors) is of fundamental importance in order to validate the system behavior against message deadlines. This worst-case value will possibly occur in a very unlikely scenario, in which remote transmissions occur exactly at the time resulting in the worst-case interference. Therefore, it seems reasonable to also analyze the system for its average-case time behavior, and, eventually, to perform a full stochastic analysis to identify the probability associated with a given latency value for a given message.

M. Di Natale et al., *Understanding and Using the Controller Area Network Communication Protocol: Theory and Practice*, DOI 10.1007/978-1-4614-0314-2_3,

This chapter discusses methods and algorithms for performing the worst-case analysis of the latency of a message. As in many cases, the analysis is based on a system model that may be more or less representative of the reality at hand for a designer who needs to analyze a given CAN system. As for other parts of this book, we are not avoiding the issue of practical applicability, but will try to discuss the limitations of the available analysis methods and the consequences of poor design choices at length.

With reference to the architecture framework outlined in the previous chapter, we will refer to the structures and policies of the *CAN adapter* (inside the *CAN controller*) with the policies for the management of the TxObjects and RxObjects, the *CAN driver* with its queuing policies and the allocation and management policies of the TxObjects, and the *Interaction Layer*, with the mechanisms for the periodic or event-based transmission of messages resulting from the packing of application-level signals.

Starting from the early 1990s, several solutions have been derived for the worst-case latency evaluation. The analysis method, commonly known in the automotive world as Tindell's analysis (from the name of the early author [58]) has been very popular in the academic community and had a substantial impact on the development of industrial tools and automotive communication architectures. The original paper has been cited more than 200 times, its results influenced the design of on-chip CAN controllers like the Motorola msCAN and have been included in the development of the early versions of Volcano's Network Architect tool. Volvo used these tools and the timing analysis results from Tindell's theory to evaluate communication latency in several car models, including the Volvo S80, XC90, S60, V50, and S40 [19]. However, the analysis was later found to be flawed, although under quite high network load conditions, that are very unlikely to occur in practical systems. Davis and Bril provided evidence of the problem as well as a set of formulas for the exact or approximate evaluation of the worst-case message response-times (or latencies) in [21].

The real problem with the existing analysis methods, however, is a number of assumptions on the architecture of the CAN communication stack that are seldom true in practical automotive systems. These assumptions include the existence of a perfect priority-based queue at each node for the outgoing messages, the availability of one output register for each message (or preemptability of the transmit registers) and immediate (zero-time) or very fast copy of the highest priority message from the software queue used at the driver or middleware level, to the transmit register(s) or TxObjects at the source node as soon as they are made available by the transmission of a message. When these assumptions do not hold, as unfortunately is the case for many systems, the latency of the messages can be significantly larger than what is predicted by the analysis.

A relatively limited number of studies have attempted the analysis of the effect of additional priority inversion and blocking caused by limited TxObject availability at the CAN adapter. The problem has been discussed first in [60] where two peripheral chips (Intel 82527 and Philips 82C200) are compared with respect to the availability of TxObjects. The possibility of unbounded priority inversion is shown for the single

TxObject implementation with preemption and message copy times larger than the time interval that is required for the transmission of the interframe bits. The effect of multiple TxObject availability is discussed in [46] where it is shown that even when the hardware transmits the messages in the TxObjects according to the priorities (identifiers) of the messages, the availability of a second TxObject is not sufficient to bound priority inversion. It is only by adding a third TxObject, under these assumptions, that priority inversion can be avoided. However, sources of priority inversion go well beyond the limited availability of TxObjects at the bus adapter. In this chapter we review the time analysis of the ideal system configuration and later, the possible causes of priority inversion, with their relationship to implementation choices at all levels in the communication stack. We will detail and analyze all possible causes of priority inversion and describe their expected impact on message latencies. Examples derived from actual message traces will be provided to show typical occurrences of the presented cases.

3.1 Ideal Behavior and Worst-Case Response-Time Analysis

This section provides the description of the original analysis method by Tindell. Although formally not correct to address the general case, the model is valid when the worst-case response-time (or latency) of a message is not happening within a busy period (do not worry if the term is not familiar, it will be introduced soon) that extends beyond the message period (or interarrival time), and provides a very good introduction to the general analysis method. In subsequent sections we will discuss the flaw, together with the fix proposed in [21].

We assume that our CAN system is composed by a set of *periodic* messages with *queuing jitter* or *sporadic* messages. In the case of a periodic message, message instances are enqueued at periodic time instants. For sporadic messages, two consecutive instances of the same message are enqueued with a *minimum interarrival time*. In the case of periodic arrival of message instances in a stream, the actual message queuing is performed by a task (denoted here as TxTask) inside the Interaction Layer. Given that this task is subject to variable response-times because of the CPU scheduling and possibly also of variations in its own computation time, the time instants at which messages are actually queued can be delayed with respect to the ideal reference periodic event sequence. This delay can be modeled as a variable with a known upper bound, defined as *queuing jitter*.

For the analysis model, we will denote a periodic or sporadic message m_i with the tuple

$$m_i = \{\mu_i, id_i, \Upsilon_i^{\mathrm{src}}, T_i, J_i, D_i\}$$

where μ_i is the length of the message in bits, id_i is the CAN identifier of the message, $\Upsilon_i^{\mathrm{src}}$ the index of the node transmitting the message, T_i is the period or the minimum interarrival time of the message, J_i is the queuing jitter (sometimes also

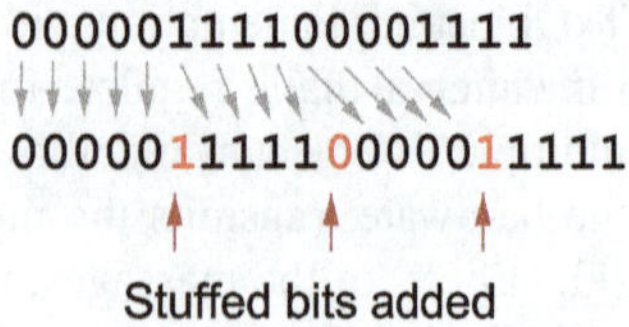

Fig. 3.1 Worst-case stuffing sequence, one bit is added for every four (except for the first five) bit in the frame

referred to as *arrival jitter*), and D_i is the deadline of the message. The definition of the queuing jitter is only significant for periodic messages. For sporadic streams we will assume that $J_i = 0$. CAN frames can only carry data packets with size in multiple of 8 bits; hence, s_i is the smallest multiple of 8 bits that is larger than the actual payload. The deadline is relative to the arrival time of the message. In almost all cases of practical interest $D_i \geq T_i$ because a new instance of a message will overwrite the data content of the old one if it has not been transmitted yet. The CAN identifier denotes the priority of the messages, and messages are indexed according to the respective identifier priority, i.e., $i < j \leftrightarrow id_i < id_j$.

Provided that a message m_i wins the contention on the bus, the transmission time e_i is given by the total number of transmitted bits divided by the bus transmission rate.

$$e_i = \frac{s_i + p + \left\lfloor \frac{s_i+34}{4} \right\rfloor}{\text{Br}}, \tag{3.1}$$

where Br is the bit rate of the transmitting CAN bus, and p is the number of protocol bits in the frame. In the case of a standard frame format (11-bit identifier) $p = 46$; for an extended frame format (29-bit identifier) $p = 65$. In the rest of the analysis, we will assume a standard frame format, but the formulae can be easily adapted to the other case. The worst-case message length needs to account for bit stuffing (only 34 of the 46 protocol bits are subject to stuffing). The number of bits that are added because of stuffing is obtained by counting the integer number of four-bit sequences in the part of the message frame subject to stuffing. It seems counterintuitive to divide by four given that a stuffing bit is added only after a sequence of five bits of the same value. However, the bit that is added because of stuffing may be (in the worst-case) the first bit of a new 5-bit sequence, hence the division by four in the formula. The situation is illustrated in Fig. 3.1.

The *queuing delay* w_i of a message m_i is the time interval from the time instant m_i is enqueued to the time it starts transmission on the bus. Note that, in reality, messages are enqueued by tasks. In most cases, it is actually a single task, conceptually at the IL, referred to as TxTask. If this is the case, then the queuing jitter of each message can be assumed to be the worst-case response-time of TxTask.

In modeling the scheduling problem for CAN, one assumption is commonly used. Each time a new bus arbitration starts, the CAN controller at each node enters the highest priority message that is available. In other words, it is assumed that the peripheral TxObjects always contain the highest priority messages in the local node,

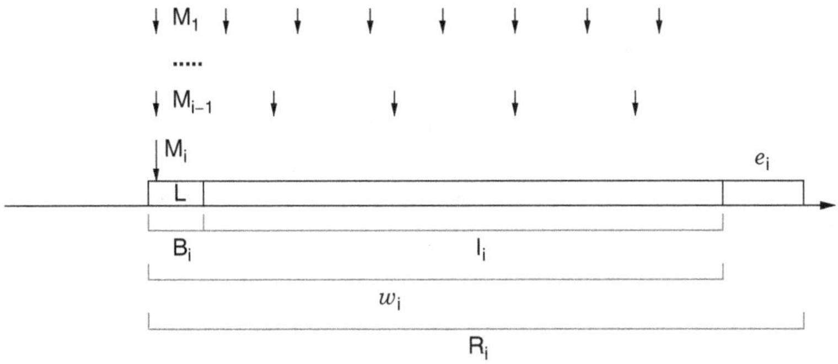

Fig. 3.2 Worst-case response-time, busy period and time critical instant for m_i

and the controller selects the highest priority (lowest id) among them. If this is the case, then the worst-case response-time R_i of message m_i is given by (Fig. 3.2).

$$R_i = J_i + w_i + e_i. \tag{3.2}$$

The queueing delay w_i consists of two factors:

- the blocking delay B_i due to a lower priority message that is transmitted when message m_i is enqueued
- the interference delay I_i due to higher priority messages that are transmitted on the bus before message m_i is transmitted

The queuing delay w_i is part of a *busy period* of level i which is defined as follows: an interval of time that starts at the time t_s when a message of priority higher than or equal to i is queued for transmission and there are no other higher priority messages waiting to be transmitted that were queued before t_s. During the busy period, only messages with priority higher than i are transmitted. The busy period ends at the earliest time t_e when the bus becomes idle or a message with priority lower than i is transmitted.

The worst-case queuing delay for message m_i occurs for some instance of m_i queued within a level-i busy period that starts immediately after the longest lower priority frame (if exists) starts its transmission on the bus. The busy period that corresponds to the worst-case response-time must occur at the critical instant for m_i [42], that is, when m_i is queued at the same time together with all the other higher priority messages in the network. Subsequently, these messages are enqueued with their highest possible rate (if sporadic). Figure 3.2 shows the situation. In effect, this requires the evaluation of the following fixed-point formula for the worst-case queuing delay:

$$w_i = B_i + \sum_{k \in hp(i)} \left\lceil \frac{w_i + J_k + \tau_{\text{bit}}}{T_k} \right\rceil e_k, \tag{3.3}$$

where $hp(i)$ is the set of messages with priorities higher than i, and τ_{bit} is the time for the transmission of one bit on the network. The solution to the fixed-point formula can be computed considering that the right hand side is a monotonic non-decreasing function of w_i. The solution can be found iteratively using the following recurrence over the integer values r, in which, at the iteration of index r, the current value of w_i is used on the right-hand side of the equation to compute the new value of w_i on the left-hand side.

$$w_i^{(r+1)} = B_i + \sum_{k \in hp(i)} \left\lceil \frac{w_i^{(r)} + J_k + \tau_{\text{bit}}}{T_k} \right\rceil e_k. \quad (3.4)$$

A suitable starting value is $w_i^{(0)} = B_i$, or $w_i^{(0)} = \epsilon$, with $\epsilon \ll e_j, \forall j$ for the lowest priority message. The recurrence relation iterates until, either $w_i^{(r+1)} > D_i$, in which case the message is not schedulable, or $w_i^{(r+1)} = w_i^{(r)}$ in which case the worst-case queuing delay is simply given by $w_i^{(r)}$.

3.2 Analysis Flaw and Correction

The flaw in the above analysis is that, given the constraint $D_i \leq T_i$, it implicitly assumes that if m_i is schedulable, then the priority level-i busy period will end at or before T_i, and the first instance of m_i is the one experiencing the worst-case latency. This assumption fails when the network is loaded to the point that the response-time of m_i is very close to its period. Given the typical load of CAN networks (seldom beyond 65%) this is a rather unusual scenario. However, when there is such a possibility, the analysis as discussed in the previous section is flawed and needs to be corrected.

The problem and the fix will be explained using a very simple bus configuration (same as the example in [21]) with three messages (A, B, and C) having a very high utilization. All messages have a transmission time of 1 unit and periods of, respectively, 2.5, 3.5, and 3.5 units (the utilization of the bus is approximately 97.1%). Suppose the priority order (the identifier assignment) is such that A is higher priority than B, and B higher than C.

Consider the analysis of message C. In this case, given that C is the lowest-priority message, there is no blocking term B. The critical instant is represented in Fig. 3.3, with all messages released at time $t = 0$. Because of the impossibility of preempting the transmission of the first instance of C, the second instance of message A is delayed, and, as a result, it pushes back the execution of the second instance of C. The result is that the worst-case response-time for message C does not occur at the first instance (three time units), but at the second one, with an actual response-time of 3.5 time units. The fix can be found by observing that the worst-case response-time is always inside the busy period for message C. Hence, to find the correct worst-case, the formula to be applied is a small modification of (3.3) that

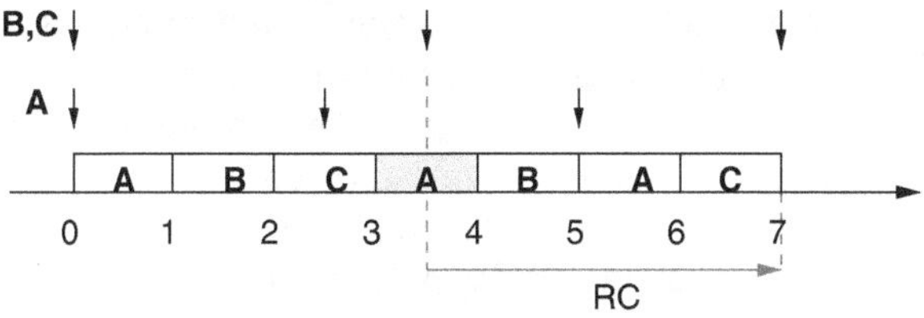

Fig. 3.3 An example showing the need for a fix in the evaluation of the worst-case response-time

checks all the q instances of message m_i inside the busy period starting from the critical instant. Analytically, the worst-case queuing delay for the qth instance in the busy period is:

$$w_i(q) = B_i + qe_i + \sum_{k \in hp(i)} \left\lceil \frac{w_i + J_k + \tau_{\text{bit}}}{T_k} \right\rceil e_i \tag{3.5}$$

and its response-time is

$$R_i(q) = J_i + w_i(q) - qT_i + e_i, \tag{3.6}$$

where q ranges from 0 to the last instance $q^{\max}$ of m_i inside the busy period. The length of the longest busy period L_i can be calculated as

$$L_i = B_i + \sum_{k=hp(i) \bigcup \{i\}} \left\lceil \frac{L_i + J_k + \tau_{\text{bit}}}{T_k} \right\rceil e_k$$

and the maximum index of the instance in the busy period is

$$q^{\max} = \left\lceil \frac{L_i + J_i}{T_i} \right\rceil - 1.$$

After the end of the busy period, the worst-case is computed as

$$R_i = \max_{q=0,\ldots,q^{\max}} \{R_i(q)\}. \tag{3.7}$$

Alternatively, an upper bound to the worst-case response-time may simply be found by substituting for all messages the worst-case frame transmission time in place of B_i in (3.5).

Note that the previous analysis methods are based on the assumption that the highest priority active message at each node is considered for bus arbitration. This seemingly simple statement, backed by intuition and common sense, is not as guaranteed as it may seem and indeed is often not true in practical systems. The failure of this assumption has many implications that will be discussed next. Before getting into details, the definition of *priority inversion*, as opposed to *blocking*, must be discussed. *Blocking* is defined as the amount of time a message m_i must wait because of the ongoing transmission of a lower priority message on the network (at the time m_i is queued). Blocking cannot be avoided and derives from the impossibility of preempting an ongoing transmission. *Priority inversion* is defined as the amount of time m_i must wait for the transmission of lower priority messages because of other causes, including queuing policies, active waits and/or other events. Priority inversion may occur after the message has been queued.

3.3 Analysis of Message Systems With Offsets

The analysis of the previous section considers the case of so-called *offset-free* systems, that is, CAN message systems in which the arrival of each periodic message stream is not bound in any way. Any two periodic streams can have an arbitrary offset and the worst-case analysis identifies the so-called *critical instant* (all higher priority messages are queued at the same time), as the beginning of the longest busy period.

In practice, however, a set of messages does not have to be *offset-free*. Even when there is no clock synchronization among nodes, the queuing of messages on each node is typically performed by a single TxTask in the *Interaction Layer*. Of course, there can also be the case in which nodes are synchronized and a queuing offset among pair of streams can be enforced.

In this case, the offset-free analysis can be pessimistic, since the offset configuration can prevent the global critical instant from ever happening. Consider, for example, the message set of Fig. 3.4, in which two messages, m_i and m_j, with periods $T_i = 4$ and $T_j = 6$ are queued with relative offset $O_{ij} = 1$ in a system in which all other messages are offset free. Clearly, m_i and m_j are never queued at the same time. In this case, the worst-case analysis requires that all busy periods starting at any of the message queuing times (between m_i and m_j) in the application hyperperiod (a time interval with length equal to the least common multiple of the message periods) are considered as candidates for finding the worst-case

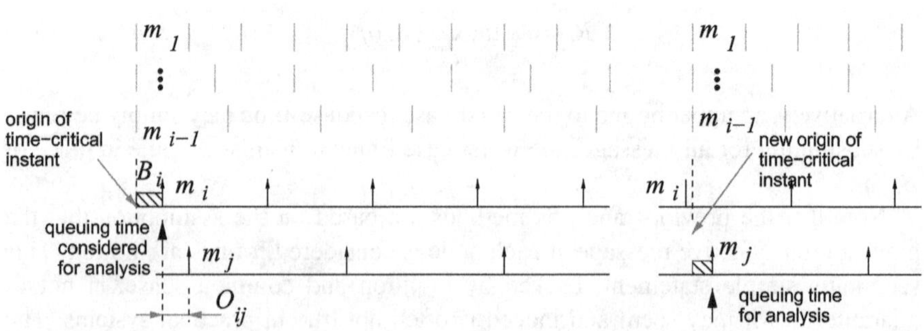

Fig. 3.4 In systems with offsets the analysis must be repeated for all the message queuing times inside the system hyperperiod

message latency. Hence, as shown in the figure, the analysis will need to consider the first queuing time for m_i, with all the offset-free messages sent according to the time critical instant rule. Then, the analysis will consider the first queuing of m_j, changing the transmission scenario for the offset-free messages in such a way that a new critical instant is constructed for the first queuing of m_j and so on.

3.4 Message Buffering Inside the Peripheral

The configuration and management of the peripheral transmit and receive objects is of utmost importance in the evaluation of the priority inversion at the adapter and of the worst-case blocking times for real-time messages. A set of CAN controllers is considered here with respect to the availability of message objects, priority ordering (when multiple messages are ready for transmissions inside TxObjects), and the possibility of aborting a transmission request and changing the content of a TxObject. The last option can be used when a higher priority message becomes ready, and all the TxObjects are used by the lower priority messages. There is a large number of CAN controllers available on the market. For the purpose of illustrating the issues, seven controller types from major chip manufacturers are analyzed. The results are summarized in Table 3.1. The chips listed in the table are Microcontroller Units (MCUs) with integrated CAN controllers or simple bus controllers. In case both options are available, controller product codes are shown between parenthesis. All controllers allow polling-based and interrupt-based management of both transmission and reception of messages.

Table 3.1 Summary of properties for some existing CAN controllers

Model	Type	Buffer type	Priority and abort
Microchip MCP2515	Standalone controller	2 RX–3 TX	Lowest message ID, abort signal
ATMEL AT89C51CC03 AT90CAN32/64	8 bit MCU w. CAN controller	15 TX/RX Msg. objects	Lowest message ID, abort signal
FUJITSU MB90385/90387 90V495	16 bit MCU w. CAN controller	8 TX/RX Msg. objects	Lowest TxObject num. Abort signal
FUJITSU 90390	16 bit micro w. CAN controller	16 TX/RX Msg. objects	Lowest TxObject num. Abort signal
Intel 87C196 (82527)	16 bit MCU w. CAN controller	14 TX/RX + 1 RX Msg. objects	Lowest TxObject num. Abort possible (?)
INFINEON XC161CJ/167 (82C900)	16 bit MCU w. CAN controller	32 TX/RX Msg. objects (2 buses)	Lowest TxObject num. Abort signal
PHILIPS 8xC592 (SJA1000)	8 bit MCU w. CAN controller	One TxObject	Abort signal

Some of these chips have a fixed number of TxObjects and RxObjects; others give the programmer freedom in allocating the number of objects available for the role of transmission or reception registers. When multiple messages are available in the TxObjects at the adapter, most chips select for transmission the data register which comes first in the controller register bank (the one with the lowest hardware identifier), *not necessarily the one containing the message with the lowest Id.* Ideally, in most chips, a message that is copied inside a TxObject can be evicted from it (i.e., the object is preempted or the transmission is aborted), unless the transmission is actually taking place. In the following sections we will see how these decisions can affect the timing predictability, and become a possible source of priority inversion.

3.5 An Ideal Implementation

The requirement that among the messages ready for transmission, the one with lowest identifier is selected at each node for the next arbitration round on the bus can be satisfied in several ways. The simplest solution is that the CAN controller has enough TxObjects to accommodate all the outgoing message streams. The situation is shown in Fig. 3.5. Even if the controller transmits messages in order of TxObject identifier, it is sufficient to map messages to objects in such a way that the (CAN identifier) priority order is preserved by the mapping.

This solution is possible in some cases. In [60] it is argued that, since the total number of available objects for transmission and reception of messages can be as high as 14 (for the Intel 82527) or 32 (for most TouCAN-based devices), the device driver can assign a buffer to each outgoing stream and preserve the

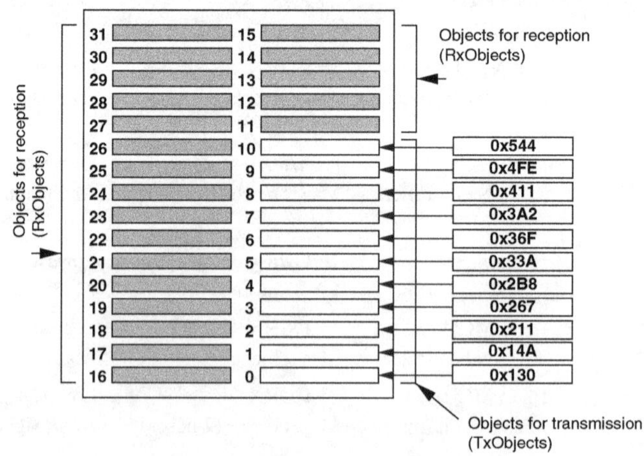

Fig. 3.5 Static allocation of TxObjects

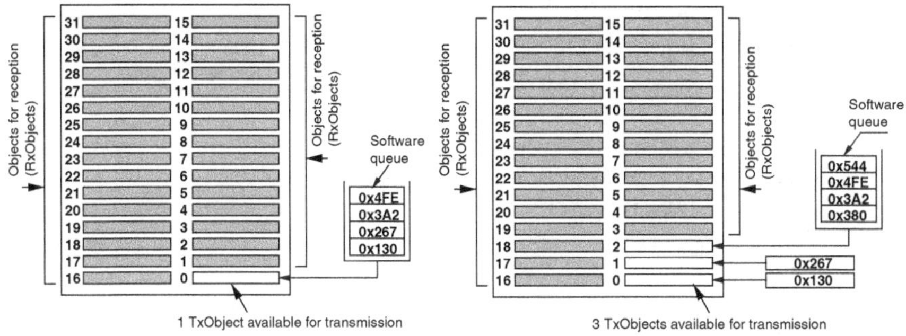

Fig. 3.6 Temporary queuing of outgoing messages

identifier order in the assignment of buffer numbers. Unfortunately, this is not always possible in several automotive architectures where message input is typically polling-based rather than interrupt-based and a relatively large number of buffers must be reserved to input streams in order to avoid message loss by overwriting. Furthermore, the number of outgoing streams can be very large for some ECUs, for example, gateway ECUs. Of course, in the development of automotive embedded solutions, the selection of the controller chip is not always an option and designers should be ready to deal with all possible hardware configurations.

The other possible solution (for example available in the CAN drivers by Vector), when the number of available TxObjects is not sufficient, is to put a subset of (or possibly all) the outgoing messages in a software queue (Fig. 3.6) as a temporary storage for accessing one or more TxObject currently used by other messages. When a TxObject is available, a message is extracted from the queue and copied into it. In this case, the preservation of the priority order of the messages in the access to the bus entails the following:

- The contents of the software queue must be kept sorted by message priority (i.e., by message CAN Identifier).
- When a TxObject becomes free, the highest priority message in the queue is immediately extracted and copied into the TxObject. This requires the interrupt-driven management of message transmissions.
- If, at any time, a new message is placed in the queue, and its priority is higher than the priority of any message in the TxObjects, then the lowest priority message holding a TxObject is evicted, placed back in the queue, and the newly enqueued message is copied in its place.
- Finally, messages in the TxObjects must be sent in order of their CAN Identifiers (priorities). If this conflicts with the transmission order predefined by the hardware for the TxObjects, the position of the messages in the TxObjects should be dynamically rearranged.

When any of these conditions does not hold, priority inversion occurs and *the worst-case timing analysis fails*; this implies that the actual worst-case can be larger than what is predicted by (3.5). Each of these causes of priority inversion will now

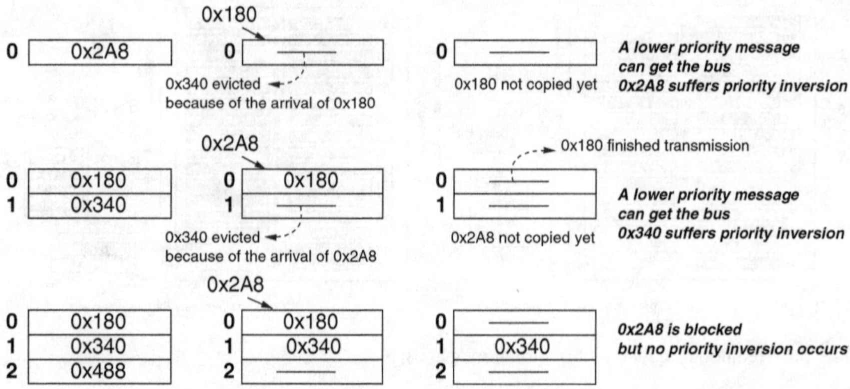

Fig. 3.7 Buffer state and possible priority inversion for the single-buffer (*top*) and two-buffer (*middle*) cases. Priority inversion can be avoided with three-buffers (*bottom*)

be analyzed in detail. However, before discussing the details, it is necessary to recall another, more subtle cause of priority inversion that may happen even when all the previous conditions are met. This problem arises because of the necessary finite copy time between the queue and the TxObjects.

3.5.0.1 Priority Inversions When Less Than Three TxObjects are Available

We will discuss two cases with possible priority inversion: single buffer with TxObject preemption, and dual buffer with preemption.

Single buffer with preemption. This scenario was first discussed in [60]. Figure 3.7 shows the possible cause of priority inversion. Suppose message `0x2A8` is in the only available TxObject when higher priority message `0x180` arrives. The transmission of `0x2A8` is aborted, and the message is evicted from the TxObject. However, after eviction, and before `0x180` is copied, a new contention can start on the bus and, possibly, a lower priority message can win, resulting in a priority inversion for `0x2A8`. The priority inversion illustrated in Fig. 3.7 can happen multiple times during the time a medium priority message (like `0x2A8` in our example) is enqueued and waits for transmission. The combination of these multiple priority inversions can result in a quite nasty worst-case scenario. Figure 3.8 shows the sequence of events that results in the worst-case delay. The top of the figure shows the mapping of messages to nodes, and the bottom part is the time-line of the message transmission on the bus and the state of the buffer. The message experiencing the priority inversion is labeled as **M**, indicating a "medium priority" message.

Suppose message **M** arrives at the queue right after message $\mathbf{L}_1$ started being transmitted on the network (from its node). In this case, **M** needs to wait for $\mathbf{L}_1$

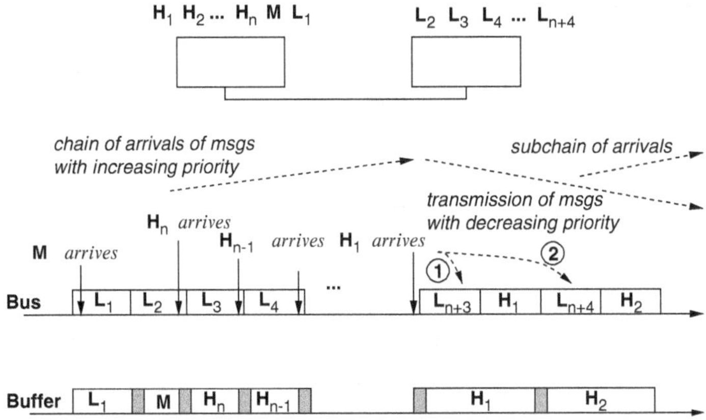

Fig. 3.8 Priority inversion for the single buffer case

to complete its transmission. This is unavoidable and considered as part of the blocking term B_i. Right after $\mathbf{L}_1$ ends its transmission, **M** starts being copied into the TxObject. If the message copy time is larger than the interframe bits, a new transmission of a lower priority message $\mathbf{L}_2$ from another node can start while **M** is being copied. Before $\mathbf{L}_2$ ends its transmission, a new higher priority message $\mathbf{H}_n$ arrives on the same node as **M** and aborts its transmission. **M** is evicted from the TxObject to be replaced by $\mathbf{H}_n$. Unfortunately, while $\mathbf{H}_n$ is being copied into the buffer, another lower priority message from another node, $\mathbf{L}_3$ can be transmitted. The priority inversion can happen multiple times, considering that the transmission of $\mathbf{H}_n$ can be aborted by another message $\mathbf{H}_{n-1}$, and so on, until the highest priority message from the node, $\mathbf{H}_1$, is written into the buffer and eventually transmitted.

The priority inversion may continue even after the transmission of $\mathbf{H}_1$. While $\mathbf{H}_2$ is being copied into the buffer, another lower priority message from another node can be transmitted on the bus and so on. At the end, each message with a priority higher than **M** allows for the transmission of two messages with priority lower than **M**, one while it is preempting the buffer and the other after transmission. Consider that during the time in which the queue of higher priority messages is emptied (in the right part of the scheduling time-line of Fig. 3.8), a new chain of message preemptions may be activated because of new arrivals of high priority messages. All these factors lead to a very pessimistic evaluation of the worst-case response-times.

Note that if message copy times are smaller than the transmission time of the interframe bits, then it is impossible for a lower priority message to start its transmission *when a message is copied right after the transmission of a higher priority message*, and the second set of priority inversions cannot possibly happen. In this case, the possibility of additional priority inversion is limited to the event of a high priority message performing preemption during the time interval in which interframe bits are transmitted. To avoid this event, it is sufficient to disable preemption from the end of the message transmission to the start of a new

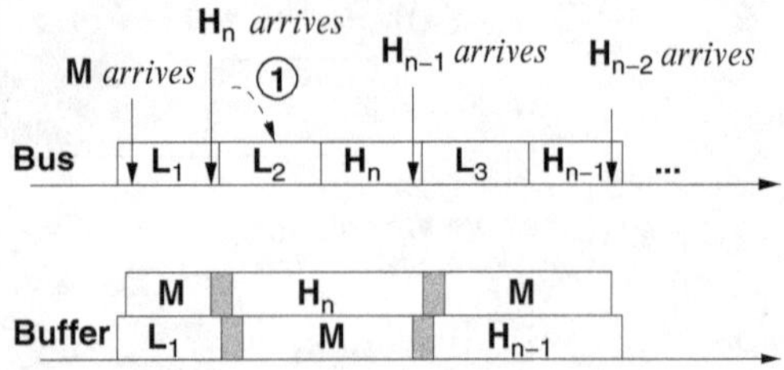

Fig. 3.9 Priority inversion for the two buffer case

contention phase. Since the message transmission is signalled by an interrupt, the implementation of such a policy should not be difficult. Under this scenario, availability of a single buffer does not prevent implementation of priority-based scheduling, and the use of the feasibility formula in [59]. The only change is that in the computation of the blocking factor B_i, the transmission time of the interframe bits must be added to the transmission time of local messages with lower priority. In the case of a single message buffer and copy times longer than the transmission time of the interframe bits, avoiding buffer preemption can improve the situation by breaking the chain of priority inversions resulting from message preemption in the first stage. However, additional priority inversion must be added considering that lower priority messages from other nodes can be transmitted in between any two higher priority message from the same node as **M**.

Dual buffer with preemption. In [46] the discussion of the case of single buffer management with preemption was extended to the case of two-buffers. Figure 3.7 shows a priority inversion event, and Fig. 3.9 shows a combination of multiple priority inversions in a worst-case scenario that can lead to large delays. The case of Fig. 3.7 defines a priority inversion when `0x340` is evicted from the TxObject while the message in the other TxObject (`0x180`) is finishing its transmission on the network. At this time, before the newly arrived message is copied into the TxObject, both buffers are empty and a lower priority message from a remote node can win the arbitration and be transmitted. The message `0x340` experiences priority inversion in this case. As an example of multiple instances of such priority inversion, the top half of Fig. 3.9 shows the scheduling on the bus (allocation of messages to nodes is the same as in Fig. 3.8), and the bottom half shows the state of the two-buffers. We assume that the peripheral hardware selects the higher priority message for transmission from any of the two-buffers.

The initial condition is the same as in the previous case. Message **M** arrives at the middleware queue right after message $\mathbf{L}_1$ started transmission and it is copied in the second available buffer. Before $\mathbf{L}_1$ completes its transmission, message $\mathbf{H}_n$ arrives. The transmitting buffer cannot be preempted and the only possible option is preempting the buffer of message **M**, evicting it from the TxObject to be replaced with $\mathbf{H}_n$. Unfortunately, during the time it takes to copy $\mathbf{H}_n$ in this buffer, the bus becomes available and a lower priority message from another node ($\mathbf{L}_2$) can perform priority inversion. This priority inversion scenario can repeat multiple times considering that a new higher priority message $\mathbf{H}_{n-1}$ can preempt $\mathbf{M}$ right before $\mathbf{H}_n$ ends its transmission, therefore allowing transmission of another lower priority message during its copy time, and so on.

The conclusion discussed in [46] is that the only way to avoid having no buffer available at the time a new contention starts, which is ultimately the cause of priority inversion from lower priority messages, is to have at least three-buffers available at the peripheral which are sorted for transmission priority according to the priority of the messages contained in them.

3.6 Sources of Priority Inversion: When the Analysis Fails

Next, we will provide a very quick glance to several other driver configuration issues or even controller policies that can lead to (possibly multiple) priority inversion and ultimately cause the worst-case latency analysis formula to fail.

3.6.1 Message Queue Not Sorted by Priority

Using FIFO queuing for messages inside the CAN driver/middleware layers may seem an attractive solution, because of its simplicity and the illusion that a faster queue management improves the performance of the system. When the message queue is FIFO, preemption of the TxObjects makes very little sense. In this case, a high priority message that is enqueued after lower priority messages will wait for the transmission of all the messages in front of it. Any message in the queue will add to its latency the transmission latency of the messages enqueued before it, with a possible substantial priority inversion.

However, there is a case where priority inversion will not occur. This may happen when messages are enqueued by a single TxTask, the TxTask puts them in the FIFO queue in priority order, and the queue is always emptied before the next activation of the TxTask. The latter condition may be unlikely, considering that the activation period of TxTask is typically the greatest common divider of the message periods.

3.6.2 Messages Sent by Index of the Corresponding TxObject

For the purpose of worst-case timing analysis, controllers like those from Microchip [47] and ATMEL [14] exhibit a desirable behavior. These chips provide at least three transmission buffers, and the peripheral hardware selects the buffer containing the message with the lowest ID (the highest priority message) for attempting a transmission whenever multiple buffers are available. Other chips, from Fujitsu [28],

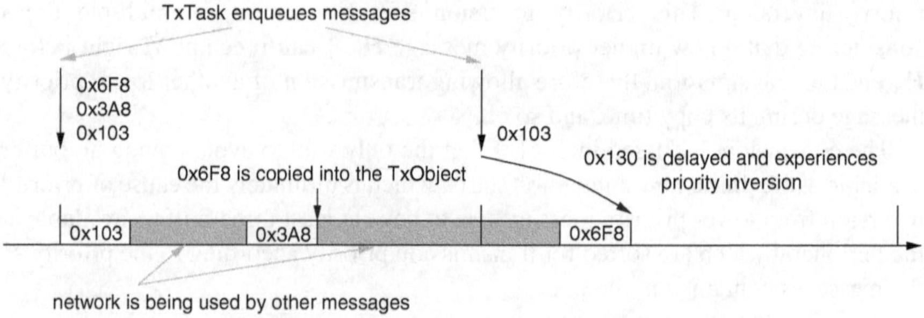

Fig. 3.10 Priority inversion when the TxObject cannot be revoked

Intel [33], and Infineon [32], provide multiple message buffers, but the chip selects the lowest (hardware) buffer index for transmission (not necessarily the one containing the message with the lowest CAN identifier) whenever multiple buffers are available. Finally, the Philips SJA1000 chip [55], an evolution of the 82C200 discussed in [60], still retains the limitations of its predecessor, that is, a single output buffer. When TxObjects are not preempted, the CAN controller may, at least temporarily, subvert the priority order of CAN messages. If, at some point, a higher priority message is copied in a higher id TxObject, it will have to wait for the transmission of the lower priority messages in the lower id TxObjects, thereby inheriting their latency (the behavior is equivalent to a FIFO queue of length one). This type of priority inversion is however unlikely, and restricted to the case in which there is dynamic allocation of TxObjects to messages, as is the case when a message extracted from the message queue can be copied in a set of TxObjects. In almost all cases of practical interest, there is either a one-to-one mapping of messages to TxObjects, or a message queue associated with a single TxObject.

3.6.3 Impossibility of Preempting the TxObjects

A special case consists of a single queue sorted by priority, where the messages extracted from the queue use a single TxObject. In addition, the TxObject cannot be preempted, that is, when a message is copied into it, the other messages in the queue need to wait for its transmission. Also in this case, the behavior is the same as that of a single-position FIFO queue (the TxObject). All the messages in the priority queue may be blocked by a possible lower priority message waiting for transmission in the TxObject.

A typical scenario that considers enqueuing of messages by a TxTask is shown in Fig. 3.10. In this case, there is a possible priority inversion when a lower priority message (`0x6F8` in the figure), enqueued by a previous instance of the TxTask, is still waiting for transmission in the TxObject when the next instance of TxTask arrives and enqueues higher priority messages (like `0x103` in the figure).

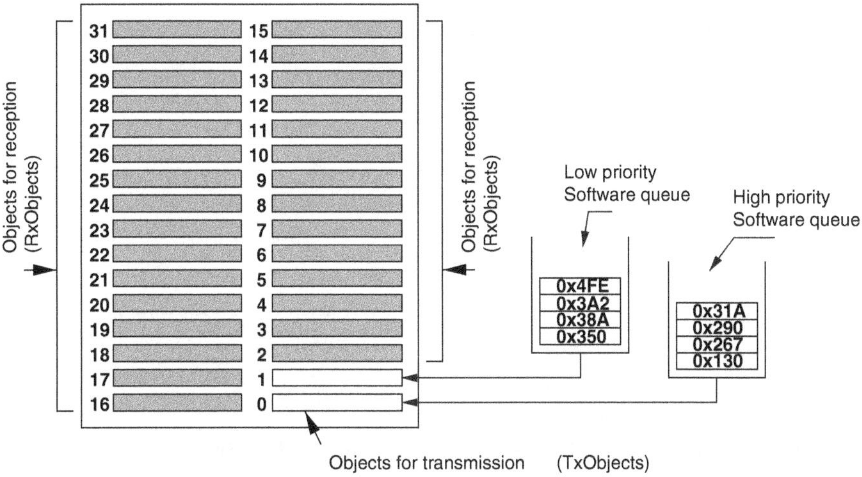

Fig. 3.11 Double software queue

Fig. 3.12 Trace of a priority inversion for a double software queue

This type of priority inversion clearly violates the rules on which (3.5) is derived. Determination of the worst-case response-time of messages becomes extremely difficult because of the overly complicated critical instant configuration for this case. A way of separating concerns and alleviating the problem could be to separate the outgoing messages in two queues: a high priority queue, linked to the highest priority (lowest id) TxObject, and a lower priority queue, linked to a lower priority object (Fig. 3.11).

Besides the possibility that the same type of priority inversion described in Fig. 3.10 still occurs for one or both queues, there is also the possibility that a slight priority inversion between the messages in the two queues occurs because of finite copy times between the queues and the TxObjects. In this case, a variation of the single-available TxObject case occurs. Figure 3.12 shows a CAN trace where a message `0x20A` from the lower priority queue manages to get in between two messages from the higher priority queue. `0x20A` takes advantage of the copy time from the end of the transmission of message `0x138` to the copy of the following message `0x157` from the high priority queue to the TxObject. In this case (not uncommon), the copy time is larger than the transmission time of the interframe bits on the bus.

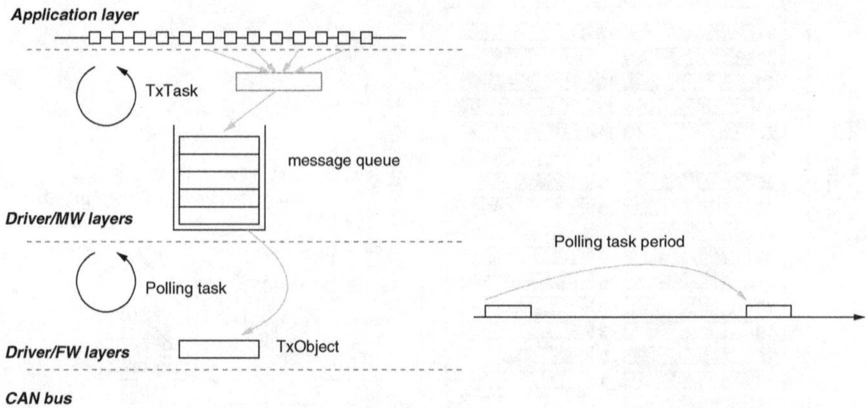

Fig. 3.13 The transmit polling task introduces delays between transmission

3.6.4 Polling-Based Output

From the point of view of schedulability analysis, the case of polling-based output can be studied along the same logic as discussed in the previous examples. However, polling-based output definitely appears to be more serious than all the cases examined until now. Figure 3.13 shows the basic functioning of a polling-based output and its impact on message latency. The polling task is executed with a period T_{p}. When it executes, it checks the availability of the TxObjects. If one TxObject is free, then it extracts the (highest priority) message from the queue and copies it into the TxObject. Then it is suspended for other T_p time units. The result of this output management policy is that transmissions from a node are always separated by at least the period T_{p}. If a message is enqueued with n other messages in front of it in the queue, then it will have to wait for (at least) $(n-1)T_{\mathrm{p}}$ before being copied in the TxObject and considered for arbitration on the bus.

Figure 3.14 shows an example of an unlikely, but not even worst-case scenario. Assuming the TxObject is not evicted from a low-priority message, a message m_i must wait up to t_i^{buf} time units for the worst-case latency of a lower priority message holding the TxObject. The time interval of length t_i^{buf} starts with the copy of the lower priority message M_L on the peripheral buffer. Assuming a perfectly periodic execution of the polling task τ_{p} with negligible execution time, the start time of t_i^{buf} is also the reference time for the periodic activations of τ_{p}. After t_i^{buf}, the transmission of the lower priority message is completed, but no other message from the same node can be copied into the TxObject until the following execution of the polling task at $\lceil t_i^{\mathrm{buf}}/T_{\mathrm{p}} \rceil$. During this time interval, the bus can be idle or used by lower priority messages. For example, right before a new message is copied from the queue into the peripheral buffer, a lower priority message from another node can get the bus. This priority inversion can happen again for each higher priority message (such as M_{H1}) sent in the time interval t_i^{que}. Furthermore, the bus scheduling is now

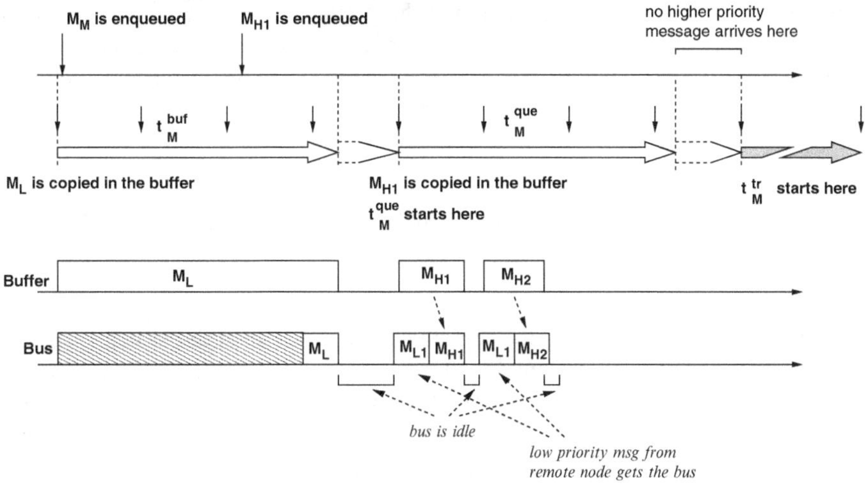

Fig. 3.14 Priority inversion for polling-based output

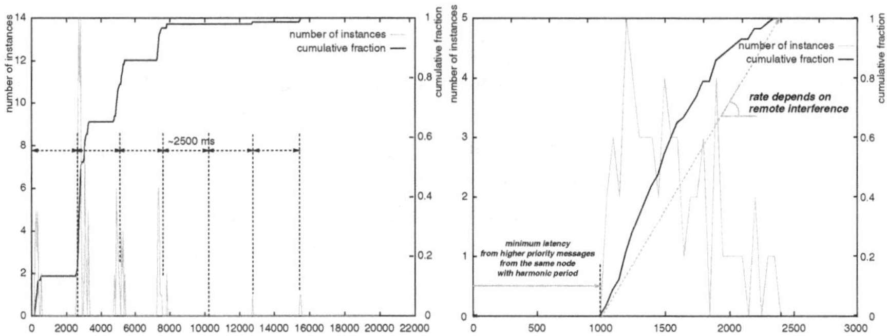

Fig. 3.15 Latency distribution for a message with polling-based output compared to interrupt-driven output

an idling algorithm, i.e., the bus can be idle even if there are pending requests in the queues of the nodes. Finding the worst-case scenario for such an idling scheme is not an easy task, and is very likely to be an NP-hard problem.

The typical result of a node with polling-based message output is shown in Fig. 3.15. The left side of the figure shows the cumulative fraction of messages (Y-axis) from a given medium priority stream that are transmitted with the latency values shown on the X-axis. Despite the medium priority, there are instances that are as late as 15 ms and more (the X-scale is in μs). This is quite unusual, compared to the typical shape of a latency distribution for a message with a similar priority, as shown in the right-hand side of Fig. 3.15. Indeed, based on its identifier, the message for which the plot was generated should not be delayed by more than a few ms. The worst-case analysis formula for this case predicted a worst-case latency of approximately 8 ms. However, the problem is that the source node uses polling-based message transmission, and the period of the polling task is 2.5 ms (as indicated by the steps in the latency distribution). The number of messages in front of it in the queue varies from 1 to 7, which is the cause of the corresponding large worst-case latency and of the substantial jitter.

3.7 From the Timing Analysis of Messages to End-to-End Response-Times Analysis

Complex distributed control systems (including today's automotive systems) consist of several Electronic Control Units (ECUs) communicating over CAN buses. CAN messages are often part of a complex, distributed application in which functions are realized by a set of distributed software tasks exchanging information over the networks. This network of tasks realizes functions that may have deadlines (hard real-time functions) or for which the performance depends on the response-times. In such a system, the timing analysis of CAN message response-times can be used late in the development stage, to ensure that the deadlines assigned to the individual messages after the functional partitioning are indeed guaranteed.

There is however, an increasing demand of timing analysis at a much earlier stage, when system architecture options must be evaluated and the (hopefully best) system configuration is selected. In this case, timing analysis could be used to evaluate a possible deployment of functions onto a candidate architecture configuration, and the evaluation of end-to-end functional delays (or response-times) is among the timing performance metrics that are of interest. Informally, the end-to-end response-time is the time that elapses from the instant when an event is detected in the environment at one end of the system, to the time instant when an actuation is commanded in response.

The analysis of the response-time in a task and message chain depends on the task and message activation model and on the way the control and data information is passed along the chain. In other words, It depends on the application model and on the software structure of the distributed system. Indeed, the software stacks provide support for the computations and communication, including application tasks, the middleware, drivers and bus peripherals.

The model of distributed real-time systems considered in this book is very simple and quite common in control systems (among others, it is supported by the AUTOSAR standard [1]). A periodic activation signal from a local clock triggers the computations on each ECU. Some application task reads the input data from a sensor, computes intermediate results that are sent over the network to other tasks and, finally, another task, executing on a remote node, generates the outputs as the result of the computation. Tasks are activated periodically and scheduled by priority. The tasks communicate through shared buffers holding the freshest data values until overwritten by new data; this model of information sharing is referred to as *communication by sampling*.

Communication between tasks takes place at the same abstraction layer, but also across layers, e.g., between application tasks and the Interaction Layer transmit task (TxTask) that is activated periodically and has the responsibility of assembling each message from the signal values and then enqueuing it for transmission on the bus. If the IL task responsible for message queuing has highest priority and completes in a very short time, message queuing can be assumed to be periodic and information is propagated by sampling across a set of scheduled objects (tasks and messages) activated periodically.

In an abstract model, end-to-end computations occur according to a dataflow. We restrict the dataflow to be acyclic, thus the system is a Directed Acyclic Graph (DAG) defined as a tuple (V, E), where V is the set of vertices representing the *tasks* and *messages* in the system, and E is the set of edges representing the *data signals* communicated among them.

$V = \{o_1, \ldots, o_n\}$ is the set of objects implementing the computation and communication functions of the system, i.e., tasks and messages are considered equally for the purpose of the communication dataflow. Task objects are also denoted as τ_i, and message objects as m_i. A *path* from o_i to o_j, denoted as $\Pi_{i,j}$, is an ordered sequence $(o_i, \ldots, o_j)$ of objects such that there is an edge between any two consecutive objects. Figure 3.16 gives an example of a path between the objects o_1 and o_7, where the figure on the top left is the architecture of the resources (CAN bus and ECUs) and objects. The activation of the source object (o_1) represents the detection of an external event, and the completion of the sink object (o_7) represents the actuation output. Resources Υ_1 and Υ_3 are ECUs, Υ_2 is a CAN bus.

Each path consists of one or more local and remote interactions. In Fig. 3.16, local interactions are between tasks o_2 and o_3, and in the chain $o_5 \rightarrow o_6 \rightarrow o_7$. Each message is enqueued by an IL task at the transmitting node. As such, the interaction between the transmitting task and the message is part of the local chain, like $o_3 \rightarrow o_4$ in the figure. Other interactions between a message and the receiving task, such as the edge (o_4, o_5) where the input and output objects are activated according to two unsynchronized clocks are *remote*. If the period of the reader o_k is larger than the period of o_h (under-sampling), then some data values sent by o_h may be overwritten before they are read, and are never propagated. In the case of over-sampling ($T_k < T_h$), some data values are read by multiple instances of o_k.

Consider the communication chain $o_2 \rightarrow o_3 \rightarrow o_4$ in Fig. 3.17. The data produced by the job $\Gamma_{2,i}$ of o_2 is never propagated because the next instance $\Gamma_{2,i+1}$ finishes execution and overwrites it before the instance $\Gamma_{3,j}$ of the receiving task is activated. On the other hand, the data generated by $\Gamma_{3,j}$ is read and propagated by two instances $M_{4,k}$ and $M_{4,k+1}$ of the following message o_4. For the path $\Pi_{i,j} = (o_i, \ldots, o_j)$, we are interested in the time interval required for an input at the source task o_i of the chain to be propagated to the last task o_j at the other end of the chain. If intermediate results are overwritten before they are read, the corresponding dataflows are not propagated (like the datapath ending with the output of $\Gamma_{2,i}$ in Fig. 3.17). Also, if some data is read multiple times, then all the response-time values associated with this data item should be considered. In the figure, this

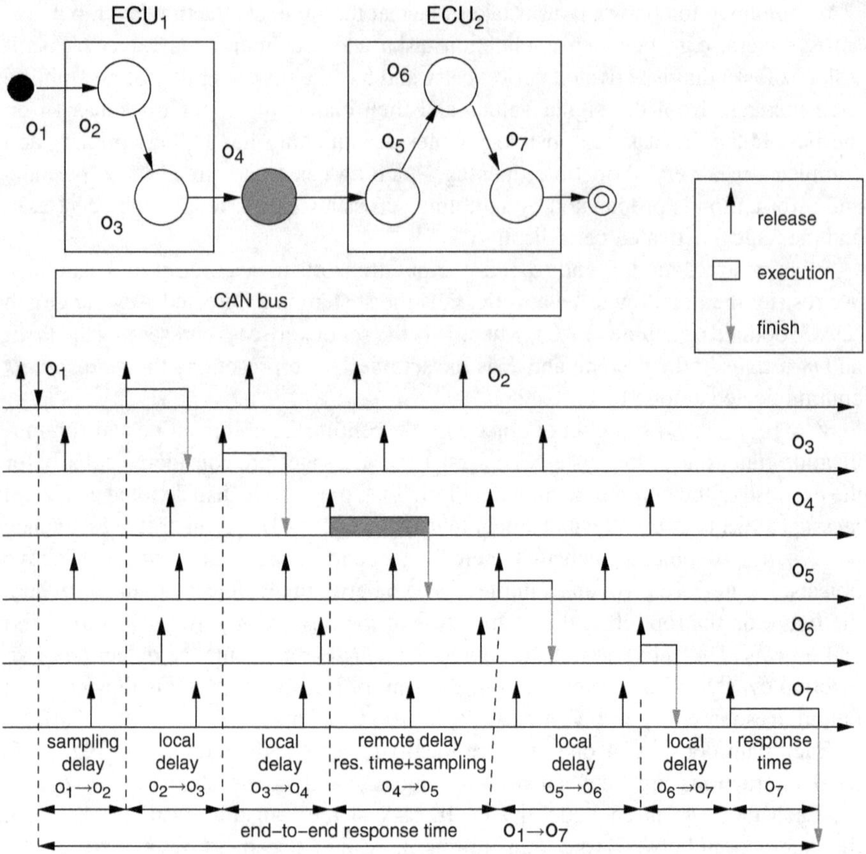

Fig. 3.16 Model of a distributed computation and its end-to-end latency

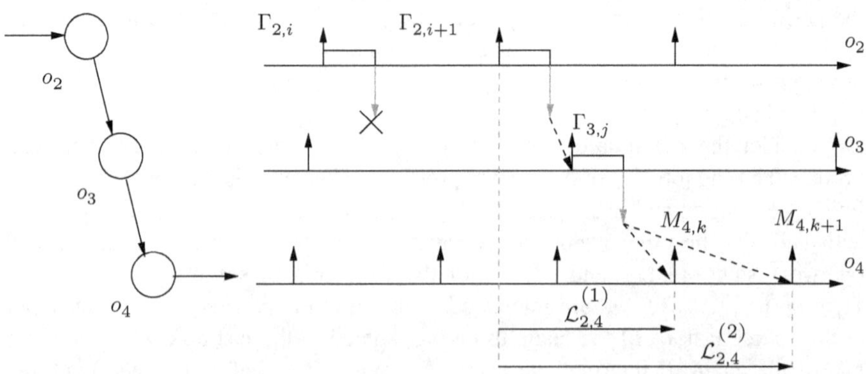

Fig. 3.17 Data loss and duplication during communication

happens for the two response-time values $\mathcal{L}_{2,4}^{(1)}$ and $\mathcal{L}_{2,4}^{(2)}$. The *end-to-end response-time* $\mathcal{L}_{i,j}$ of path $\Pi_{i,j}$ can be defined as the interval between the activation of one instance of o_i and the completion of the instance of o_j that produces a result dependent on the output of o_i (but other definitions are possible, depending on the measure of interest).

In conclusion, the evaluation of the worst-case end-to-end response-time for a given path (see Figs. 3.16 and 3.17) is the result of the composition of task and message response-times, as required to produce the result of the computations or to deliver data signal from one ECU to another, but also sampling delays, which in turn depend on the periods of the messages and the tasks and possibly on their activation offsets. For a detailed discussion of the end-to-end analysis, the reader can refer to [65].

3.8 Conclusions

You may feel a little bit confused at the end of this chapter, and you probably should. Besides technicalities, formulas and the description of unlikely (and seemingly overcomplicated) scenarios, the final lesson of this chapter might be read as follows: "Here is a nice formula for predicting the worst-case response-time of messages, the only problem is… in many cases it will not work!". Partly (and sadly) this is true. However, there is another possible way of reading this chapter, this time with the eye of the designer. Here the question is: "How should the CAN driver and interaction layers be designed, defined and possibly developed in such a way that (multiple) priority inversions can be avoided and the system is predictable?". Designers are often faced with these questions: "Should I use a SW queue? How many queues? Is it acceptable to have FIFO queuing of messages? What is the cost of such a queue?" Finally, even in those cases it is formally not correct, the analysis formula can still provide an estimate of the worst-case time performance of messages. It could be used then to predict the consequences of the assignment of an identifier to a message, to define at design time what is the best system configuration as compared to other possible options.

Chapter 4
Stochastic Analysis

4.1 Introduction

This chapter focuses on the stochastic analysis of the timing performance of CAN messages. Worst-case analysis based on schedulability theory allows to verify the timing correctness of a CAN subsystem. However, as discussed in the previous chapter, CAN messages carry the signal data realizing communication in (possibly complex) end-to-end functionality, in which the flow of data and control may go across several nodes and networks. In this case, timing analysis can be used to compute the contribution of messages to end-to-end latencies and provides the architecture designer with a set of values (one for each end-to-end path) on which he/she can check correctness of an architecture solution.

An example of such functions (and the original motivation for the analysis presented in this chapter) are *active safety functions* in automotive systems. These functions gather a 360° view of the environment via radars and cameras, and require several processing stages before the actuation signals are produced, including sensor fusion, object detection, control and arbitration layers. Examples are Adaptive Cruise Control (ACC) or Lane Keeping Systems. In an adaptive cruise control system, a set of radars (or other sensors) scans the road in front of the car to ensure that there are no other cars or objects moving at a lower speed or even stopped in the middle of the lane. If such an object is detected, the system lowers the target speed of the cruise control until it matches the speed of the detected obstacle. A hard deadline can be defined for such a system (the worst-case reaction time that allows preventing a collision, as discussed in the last chapter), but clearly a faster average reaction time is always preferable and the designer can leverage additional knowledge on the probability distribution of the function response-times. Worst-case analysis should be complemented by probabilistic analysis for two main reasons:

- Many applications are not time-critical, but the performance of the controls depend on the average response-time, which needs to be analyzed and minimized. This is true also for many time-critical functions in which, among all the solutions

M. Di Natale et al., *Understanding and Using the Controller Area Network Communication Protocol: Theory and Practice*, DOI 10.1007/978-1-4614-0314-2_4,

that guarantee the deadlines (therefore, formally correct), some may perform better than others. One class of functions for which information on the average response-times is a fundamental parameter for evaluating the quality of the architecture solution are the aforementioned *active safety* functions. Among those are the *Lane Keeping*, *Lane Departure Warning*, *Automatic Cruise Control* and also the *Crash Preparation*. Other functions that are similarly affected by the average response-time behavior are driving assistance functions like the *Assisted or Automated Parking*. The latter (the car is automatically parked by the control system in a free spot detected by radars or other sensors) is a clear example of an application in which there is clearly no deadline, but the quality of the service perceived by the user depends nevertheless on the time required by the function to complete. However, for various reasons, even the primary functions of *active safety* systems (not including the diagnosis functions) are designed and developed without hard deadlines. These systems are intended to assist the driver, who always remains in control. A shorter response-time may facilitate earlier activation of the system, and may thereby enhance the system effectiveness. For example, in the case of a Lane Departure Warning, under some circumstances an earlier warning that the car is inadvertently departing from the driving lane may allow more time for the driver to react and recover.

- In the periodic activation model with communication by sampling (see the end of the previous chapter), each time a message is transmitted or received, a task (message) may need to wait up to an entire period of *sampling delay* to read (forward) the latest data stored in the communication buffers. Adding worst-case delays at each step allows to obtain the worst-case latencies of paths, but the probability of such a worst-case value can be very small. So small in fact (in several functional paths examined in experimental vehicles it is less than 10^{-14} with a period of 50 ms, i.e., 7.2×10^{-10}/hour), to be smaller than the probability of HW failure (as a reference, for the international safety standard ISO 26262, the most stringent Automotive Safety Integrity Level—ASIL D—has a targeted random hardware failure of 10^{-9}–10^{-8} faults/hour [2])! In this case, designing for worst-case time behavior can be wasteful.

In this chapter, we present a stochastic analysis method to compute the probability distribution of CAN message response-times. This method is an alternative to deriving probability distributions from (possibly expensive and extensive) simulations and statistical modeling. The message sizes and activation periods are assumed to be known. Also, the distributions of the random transmission times due to the non-deterministic bit stuffing are computed assuming a random data content. The system model assumed by the analysis is that of non-deterministic message queuing times with unsynchronized ECUs, consistent with the architecture framework described in the previous chapters and with the industrial practice. In most CAN driver and middleware-level implementations, a single interaction layer task is responsible for queuing all messages originating from a node.

The following section provides an overview of the recent work in this area. In Sect. 4.3 the formal model used for the stochastic analysis is defined, and Sect. 4.4

presents the stochastic analysis framework that approximates the CAN message response-times. The analysis results are demonstrated on an example automotive CAN system, which provides a good approximation of the latency distribution compared with the results obtained by simulations and trace data.

4.2 Reference Work

Probabilistic analysis of priority-scheduled systems has been addressed in the scientific literature. Gardner [29] computes the probability of deadline misses for each job released in the busy interval, and chooses the maximum as an upper bound on the probability of deadline misses for the corresponding task. Simulation-based analysis and stochastic methods exist [23, 43] for computing the probability density functions (pdfs or probability mass functions pmfs in the discrete case) of the response-times of tasks scheduled on single or multi-processor platforms. For messages scheduled on a CAN bus, Navet et al. [49] introduce the concept of worst-case deadline failure probability (WCDFP) because of transmission errors. WCDFPs are computed with respect to the *critical instant*, i.e., the worst-case response-time scenario. In Nolte et al. [50] the worst-case response-time analysis is extended using random message transmission times that take into account the probability of a given number of stuff bits, but the analysis is still performed in the worst-case scenario. Worst-case analysis of task and message response-times in priority-based systems addressed independent periodic tasks [39], tasks with offsets and jitter [53], and OSEK systems [41]. Davis et al. [22] discuss scheduling and response-time analysis of CAN messages.

In Lehoczky's Real-Time Queueing Theory [40], the task set is modeled as a Markovian process. Given a scheduling algorithm and the distribution of the deadlines, the distributions of the times remaining until the deadline of jobs are computed. The analysis is performed in heavy traffic conditions, when the behavior can be approximated by a diffusion process. Arrival and execution times are modeled by Poisson distributions, and the system has a very large set of tasks with very high utilization ($U \rightarrow 1$). In an independent but related work, Kim and Shin [36] model an application as a queuing network scheduled as first-in first-out (FIFO), and use exponentially distributed task execution times. The underlying mathematical model is a continuous-time Markov chain.

Manolache presents an analytical method to compute the expected deadline-miss ratio for single processor systems, and an approximate method for multi-processor systems [44].

The most relevant method to the approach presented in this chapter was proposed by Díaz et al. [23, 43], in which the probability mass functions (pmfs) of the response-times of a set of independent periodic tasks executed on a single processor by a preemptive priority-based scheduler is computed. The activation times of all the task instances are known and the task execution times are defined by pmfs. Díaz et al. [23] introduce the concept of P-level backlog at time t to represent the sum

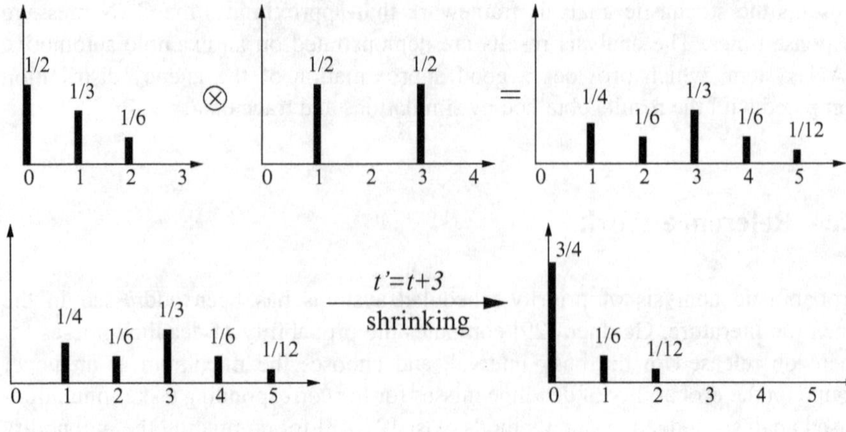

Fig. 4.1 Convolution and shrinking

of the *remaining* requested CPU times of all the task instances (jobs) of priority higher than or equal to P that have been released before t. The P-level backlog at the beginning of each hyperperiod is a random variable, and the stochastic process composed of the sequence of these backlogs is proven to be a Markov chain. When the maximum utilization is less than 1, a stationary distribution (pmf) of the backlog possible values is reached at the beginning of the second hyperperiod. Then, a stationary backlog at any time *within* the hyperperiod can be calculated using two operations called *convolution* and *shrinking*. The P-level backlog pmf right after the release of a job with priority higher than or equal to P is obtained by performing a convolution, at the job release time, of the backlog pmf with the job execution time pmf. Given the pmf of a backlog at time t, shrinking allows to compute the backlog at time $t' > t$ by shifting the pmf to the left by $t' - t$ units and by defining the probability of zero backlog be the sum of the probabilities defined for non-positive values. The time t' represents a time instant *right before the next release time of a task instance that contributes to the P-level backlog*. Indeed, shrinking is equivalent to simulating the progression of time (Fig. 4.1).

Finally, the stationary response-time pmf of a job can be computed with a sequence of splitting, convolution, and merging operations starting with the backlog at the release time and the pmf of the job execution time itself, and considering the pmfs of the execution times of higher priority jobs that are released before the job terminates, as the current job can be preempted. This process continues until either the time instant corresponding to the job deadline is reached, in which a deadline violation is found, or until the job completes its execution before any other higher priority job is released. As there may be multiple jobs of the same task released within the hyperperiod, the stationary task response-time pmf is obtained by averaging the response-time pmfs of all the jobs within the hyperperiod.

The analysis provided in [23, 43] assumes knowledge of the (relative) activation times between tasks, and it estimates blocking delays due to lack of preemption and release jitter in a pessimistic way. Other stochastic approaches assume Poisson arrivals and/or are not applicable to medium or light load conditions. Also, the stochastic analysis of CAN systems in [49] and [50] is focused on critical instant conditions, *without considering the probability of occurrence of a critical instant.* Finally, [18] considers a model of random arrival stream, but only at one priority level. In [20], a stochastic analysis framework is proposed that provides probability distributions of message response-times where the periods of messages are given by independent random variables.

4.3 System Model and Notation

The CAN arbitration protocol is both priority-based and *non-preemptive*, that is, the transmission of a message can not be preempted by higher priority messages. The protocol is described in Chap. 1 for more details, refer to the official specification in [5].

We consider a CAN system where messages are queued periodically. At each node, the bus adapter sends the queued messages in priority (ID) order and its transmit registers (TxObjects) can be overwritten if a higher priority message becomes available in the sorted queue.

A message m_i is defined by $(\Upsilon_i^{\mathrm{src}}, T_i, \mathcal{O}_i, \mathcal{E}_i, P_i)$, where $\Upsilon_i^{\mathrm{src}}$ is the ECU from which it is transmitted, T_i its period, $\mathcal{O}_i$ its initial phase, $\mathcal{E}_i$ its transmission time, and P_i its priority. The message transmission time $\mathcal{E}_i$ is a random variable[1] due to the effect of bit-stuffing, that is, when the protocol requests the addition of a bit of opposite sign on a frame containing more than five consecutive bits with the same value. The *hyperperiod* H of the system is defined as $H = \mathrm{lcm}\left(\bigcup_i T_i\right)$.

For each periodic activation, we consider a *message instance.* On its arrival, a message instance is *queued* by a periodic middleware task. The j-th instance of message m_i is denoted as M_{ij}, its arrival time as $\mathcal{A}_{ij} = \mathcal{O}_i + (j-1) \times T_i$,[2] its queuing time $\mathcal{Q}_{ij} = \mathcal{A}_{ij}$, its start time $\mathcal{S}_{ij}$, and its finishing time $\mathcal{F}_{ij}$. Note that the arrival times of the message instances from the same message are not independent, i.e., given a phase of the message, the arrival times of all its message instances are determined as well. In other words, although the initial phase is random, it is assumed to be constant in the hyperperiod.

Each message m_i is associated with a unique ID P_i, which also represents its priority. According to the CAN protocol, the lower the message ID, the higher its priority, $P_i < P_j$ implies that m_i has higher priority than m_j.

Since there is no synchronization mechanism for the local clocks of the nodes in the standard CAN protocol, from the standpoint of a message instance on node

[1] In this chapter we use calligraphic letters to denote random variables, such as E_i and O_i.

[2] The add operation of a random variable $\mathcal{V}_1$ and a normal variable V_2 is that $\forall v_1, \mathbb{P}(\mathcal{V}_1 + V_2 = v_1 + V_2) = \mathbb{P}(\mathcal{V}_1 = v_1)$.

ECU_k, the initial phase of a message m_i from any other node $\text{ECU}_l (l \neq k)$ is $\mathcal{O}_i = \Psi_i + \mathcal{O}_{lk}$, where Ψ_i is the local phase of message m_i, and $\mathcal{O}_{lk}$ the relative phase between ECU_l and ECU_k. Obviously the phases and thus the arrivals of different messages on the same node are not independent. $\mathcal{O}_{lk}$ is a real-valued random variable, modeling the random message queuing times. Without loss of generality, the pdf of the relative phase $\mathcal{O}_{lk}$ is assumed to be uniformly distributed within the interval $(0, H)$.

The *response-time* $\mathcal{R}_{ij}$ of a message instance M_{ij} is defined as the time interval from its arrival to its finish, i.e., $\mathcal{R}_{ij} = \mathcal{F}_{ij} - \mathcal{A}_{ij}$, which is a random variable that depends on the phases of the messages from other nodes, and the non-deterministic transmission times. The pmf of the response-time $f_{\mathcal{R}_i}$ of message m_i is obtained as the average of the response-time pmfs of all the message instances M_{ij} in the hyperperiod provided that the distribution is stationary.

4.4 Stochastic Analysis of Message Response-Times

In this section, we summarize a stochastic analysis framework that allows to compute message response-time probabilities with non-deterministic message phasing and captures the nature of CAN-based systems where periodic messages are sent from ECUs with unsynchronized clocks. The following sections define the steps that are required for computing the pmf of the response-time of an instance of message m_i on a generic ECU. The first step is to compute the interferences from messages on remote ECUs.

4.4.1 A Modeling Abstraction for CAN Messages

One *characteristic interference message*, or simply *characteristic message*, is used for each remote node to model the interference caused by messages from it. Messages are queued by the TxTask. Hence, every message from the same node is queued with the same phase and its period is an integer multiple of the TxTask period. From the standpoint of message m_i, the queuings of higher priority instances from a remote node ECU_k $hp(P_i)_k = \{m_j | P_j < P_i \cap \Upsilon_j^{\text{src}} = \text{ECU}_k\}$ only happen at integer multiples of the greatest common divisor (gcd) of their periods. The characteristic message summarizes the effect of all such messages, with period equal to the gcd of their periods, non-deterministic transmission time, given that it represents sets of different length queued at different time instants and, with respect to remote message m_i, a random offset, because of the lack of synchronization among ECUs.

Figure 4.2 provides an example of three remote messages with periods and transmission times equal to (60, 2), (10, 1), (20, 1), respectively. During the hyperperiod $[0, 60)$, the message instances are queued at time instants $p \times 10$, $p = 0, 1, 2, 3, 4, 5$.

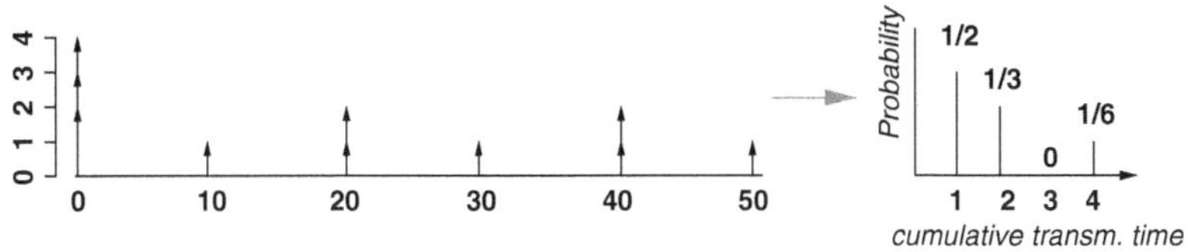

Fig. 4.2 An example of characteristic message transmission time

At time 0, all messages are queued and the total transmission time is 4. The transmission times that are requested at the other time instants are 1, 2, 1, 2, 1, respectively. From the standpoint of a remote message m_i, the first subset that can produce interference is one of the six at random, given that the ECU clock offsets are uniformly distributed. Three time instants out of six contribute to a transmission time of 1 unit, two out of six, 2 units, and in one case out of six, 4 units. Thus the pmf of the characteristic message transmission time is defined as $\mathbb{P}(1) = 1/2, \mathbb{P}(2) = 1/3, \mathbb{P}(4) = 1/6$.

Different from the example in the figure, CAN message transmission times are non-deterministic because of the stuff bits. A simple model for computing the probability of a given number of stuff bits can be obtained by assuming equal probability of a zero or one value for each bit. Other models and methods can be found in [51].

Algorithm 1 computes the transmission time pmf $f_{\mathcal{E}_c}$ of the P_i-level characteristic message on a generic ECU_k. First, it identifies the set of messages $hp(P_i)_k$, the lcm of their periods H_k^i, and the gcd of their periods T_c, the time interval that separates their queuing instants. Then, for each multiple of T_c within $[0, H_k^i)$, i.e., $t = pT_c, p = 0, 1, \ldots, H_k^i/T_c - 1$, the pmf of the total transmission time $\mathcal{E}[p]$ is calculated by adding (by convolution) the transmission time pmfs of the message instances queued at t. Finally, $f_{\mathcal{E}_c}$ is the average of the pmfs of $\mathcal{E}[p]$ at these time instants.

The characteristic message introduces inaccuracy in the analysis. If the response-time of any instance of m_i is smaller than the minimum among the T_c of all remote nodes, then it will suffer interference from at most one set of message instances from any other node. Such set will be random, because of the random offsets among node clocks. Similarly, interference from the characteristic message will occur at most once and the amount of interference is the random transmission time of the characteristic message, which is the same as the interference from a random set of message instances for one of the pT_c time instants in the real case. However, when more than one interference is possible, in reality there is a correlation between the amount of interference provided by instances queued at consecutive cycles of the TxTask, which is lost in the characteristic message. In the example of Fig. 4.2, an interference of 4 units is *always* followed by an interference of one unit. With the characteristic message, there is a 1/36 chance that there will be another interference of 4 units. In general, even if the condition for absence of errors (response-time of

Algorithm 1 Calculate the transmission time pmf $f_{\mathcal{E}_c}$ of P_i-level characteristic message m_c for ECU_k

$hp(P_i)_k = \left\{m_j \mid P_j < P_i \cap \Upsilon_j^{\text{src}} = \text{ECU}_k\right\}$
$H_k^i = \text{lcm}\left(\bigcup_{j \in hp(P_i)_k} T_j\right),\ T_c = \text{gcd}\left(\bigcup_{j \in hp(P_i)_k} T_j\right)$
for $p = 0, 1, \ldots, H_k^i / T_c - 1$ **do**
  $f_{\mathcal{E}[p]}(0) = 1.0$
  $t = p \times T_c$
  for each $j \in hp(P_i)_k$ **do**
    if $t \bmod T_j = 0$ **then**
      $f_{\mathcal{E}[p]} = f_{\mathcal{E}[p]} \otimes f_{\mathcal{E}_j}$
    end if
  end for
end for
for $e = 0$ to $E_c^{\max}$ step τ **do**
  $f_{\mathcal{E}_c}(e) = \dfrac{T_c}{H_k^i} \sum_{p=0}^{H_k^i/T_c - 1} f_{\mathcal{E}[p]}(e)$
end for

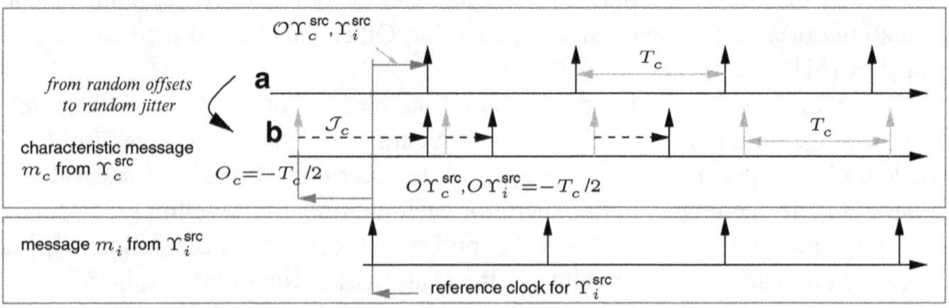

Fig. 4.3 The transformation of the queuing model for remote messages

m_i always less than $\min\{T_c\}$) is true only for a (high priority) subset of all messages, the method retains sufficient accuracy for the evaluation of message response-times even when this assumption is violated, as the experiments show.

The characteristic message allows a significant reduction in the size of the space to be analyzed. First, only one message needs to be considered for each remote ECU. In addition, with respect to a remote message m_i, the only significant offset values of a characteristic message are in the range $[0, T_c)$, which is significantly smaller than $[0, H_k^i)$.

The second approximation/simplification in the model consists in changing the queuing model of remote characteristic messages from deterministic periodic with random initial offset ((a) in Fig. 4.3) into a deterministic offset, periodic activation times and random queuing jitter ((b) in Fig. 4.3).

The characteristic message m_c is queued with a random offset with respect to m_i which depends on $\mathcal{O}_{\Upsilon_c^{src},\Upsilon_i^{src}}$ (we assume $O_c = 0$ and $O_i = 0$). In the approximate model, $\mathcal{O}_{\Upsilon_c^{src},\Upsilon_i^{src}} = 0$, and the randomness in the offsets among the ECU clocks is transformed into a constant zero offset among them, and random activation times for the characteristic messages, with offset $O_c = -T_c/2$ (to ensure a single interference on m_i if its response-time is $< T_c/2$). At each activation, characteristic messages are delayed by a random jitter $\mathcal{J}_c$, uniformly distributed within the period T_c. For the jth instance $M_{c,j}$ of m_c, the queuing time is $\mathcal{Q}_{c,j} = O_c + (j-1)T_c + \mathcal{J}_c$. The initial offset $O_c = -T_c/2$ is also chosen to minimize the probability of multiple interferences from m_c to the first instance of m_i.

The introduction of random jitter in place of an initial random offset further introduces inaccuracy. In the model with random offsets, message transmissions are requested after exactly one period. This is not true for the characteristic message with a random release jitter (as shown in Fig. 4.3).

The formal definition of the *P_i-level characteristic message m_c* from $\Upsilon_c^{src} = \Upsilon_k$ with respect to m_i is

- Period $T_c = \gcd\left(\bigcup_{j \in hp(P_i)_k} T_j\right)$, where $hp(P_i)_k = \left\{m_j | P_j < P_i \cap \Upsilon_j^{src} = \Upsilon_k\right\}$
- Initial offset $O_c = -T_c/2$, $O_{\Upsilon_k,\Upsilon_i^{src}} = 0$
- Queuing jitter $\mathcal{J}_c$ uniformly distributed in $[0, T_c)$
- Random transmission time $\mathcal{E}_c$, as shown in the example in Fig. 4.2 and Algorithm 1
- Priority P_c higher than P_i

In conclusion, to analyze the response-time of a message m_i, an approximate system is defined which consists of all the messages from the same node with priority higher than P_i. These messages are not abstracted using a characterization message, since they have known queuing instants with respect to m_i. In addition, one characteristic message of level P_i is used for each of the other nodes. In the approximate system, every message m_j is modeled by $(T_j, O_j, \mathcal{J}_j, \mathcal{E}_j, P_j)$, where $\mathcal{J}_j$ is a random queueing jitter for characteristic messages and 0 for messages from the same node as m_i.

Theorem 4.4.1. *If the bus utilization is less than 1, the approximate system model based on characteristic messages and the transformation of random offsets into deterministic offsets and random jitter is stable and admits a stationary distribution solution as in [43].*

Proof. The proof is based on the results in [23, 43] where the authors demonstrate stability provided that the model is periodic and the average resource utilization is less than 1.

Demonstrating the periodicity of the system model is simple. In our case all messages are periodic, activated with known offsets and random jitter. The lcm of the message periods (or application hyperperiod) is the system period.

For the second part, the average utilization $\bar{U}_j$ of a message m_j is defined as the ratio between its expected transmission time $\bar{E}_j$ and its period T_j. First, the average utilization of a characteristic message m_c is proven to be equal to the sum of the average utilizations of the messages in the set $hp(P_i)_k$ it abstracts. Given that the execution time $\mathcal{E}_c$ of m_c is the average of the total transmission time $\mathcal{E}[p]$ at time instants $t = pT_c$, $p = 0, 1, \ldots, H_k^i/T_c - 1$ within one hyperperiod $[0, H_k^i)$, its expected value is

$$\bar{E}_c = \frac{T_c}{H_k^i} \times \sum_{p=0}^{H_k^i/T_c - 1} \bar{E}[p]. \tag{4.1}$$

Each message $m_j \in hp(P_i)_k$, is queued H_k^i/T_j times in one hyperperiod $[0, H_k^i)$. Thus, $\sum_p \mathcal{E}[p] = \sum_{j \in hp(P_i)_k} \frac{H_k^i}{T_j} \mathcal{E}_j$, and the average value is

$$\sum_p \bar{E}[p] = \sum_{j \in hp(P_i)_k} \frac{H_k^i}{T_j} \bar{E}_j = H_k^i \times \sum_{j \in hp(P_i)_k} \bar{U}_j.$$

The average utilization $\bar{U}_c$ from m_c is

$$\bar{U}_c = \frac{\bar{E}_c}{T_c} = \frac{1}{T_c} \times \frac{T_c}{H_k^i} \times \sum_p \bar{E}[p] = \sum_{j \in hp(P_i)_k} \bar{U}_j. \tag{4.2}$$

In the approximate system, when computing the latency pmf for m_i, one characteristic message of level P_i is used for each of the remote nodes. By (4.2), the utilization of each characteristic message is the sum of the utilization of the higher priority messages on its node. Hence, the average utilization of the approximate system is the same as in the original CAN system. The second part of the proof follows from the hypothesis that the original system has average utilization less than 1. □

However, the results obtained from the approximate system *are not always pessimistic* according to the definition in [43]. The analysis may lead to optimistic probabilities of deadline misses, so design decisions based on the analysis result should be taken carefully, especially for time-critical systems where missing deadlines can lead to failure of the complete system. In general, the stochastic analysis should not be a tool for system validation but rather for comparison of architecture solutions and early estimates of performance. In most experimental cases, as in Figs. 4.5 and 4.6, the plots of the analysis results are below the simulation data, meaning that the analysis gives pessimistic results (especially for long latencies). However, bounding the error in either direction is not easy.

The *P-level backlog* $\mathcal{W}_t^P$ *at time t* is the sum of the remaining transmission times of the queued message instances with priority higher than P. For each characteristic message instance $M_{p,q}$, the event "$M_{p,q}$ is queued after t" is denoted as $V_t^{p,q}$, or $V_t^{p,q} = (\mathcal{Q}_{p,q} > t)$, and its *complement* $\overline{V}_t^{p,q} = (\mathcal{Q}_{p,q} \leq t)$. Suppose t is a possible queuing time for n instances $M_{p_1,q_1}, M_{p_2,q_2}, \ldots, M_{p_n,q_n}$ with priorities higher than or equal to P. This set of instances is the *P-level message instance set at time t*, denoted as $I(P,t)$. For each instance $M_{p_k,q_k} \in I(P,t)$, the two mutually exclusive events $V_t^{p_k,q_k}$ and $\overline{V}_t^{p_k,q_k}$ divide the event space into two subspaces. When considering all instances in the set, there are totally 2^n possible combinations or event vectors, each representing a different combination of the random queuing times of the n instances with respect to t. These 2^n events represent the *P-level event space* $S(P,t) = \left\{ \left(V_t^{p_1,q_1}, V_t^{p_2,q_2}, \ldots, V_t^{p_n,q_n}\right), \left(V_t^{p_1,q_1}, V_t^{p_2,q_2}, \ldots, \overline{V}_t^{p_n,q_n}\right), \ldots, \left(\overline{V}_t^{p_1,q_1}, \overline{V}_t^{p_2,q_2}, \ldots, \overline{V}_t^{p_n,q_n}\right) \right\}$, with total probability 1 defined for each priority level P and time t.

The *queuing pattern* at time t, denoted as $\mathcal{X}_t$, is defined based on the possible queuing times of these message instances. $\mathcal{X}_t$ is a random variable defined over $S(P,t)$ with a known probability mass function. With a slight abuse of notation, $\mathcal{X}_t$ will also be used to indicate the *queuing pattern vector* at time t. For a queuing pattern $\mathcal{X}_t$, if the queuing event at t corresponding to the instance $M_{p,q}$ is $V_t^{p,q}$, then $\mathcal{X}_t$ is called a *positive queuing pattern* of $V_t^{p,q}$; otherwise, if $\overline{V}_t^{p,q}$ applies, it is a *negative queuing pattern* of $V_t^{p,q}$. The *complementary queuing pattern* $\overline{\mathcal{X}}_{t(p,q)}$ *with respect to* $M_{p,q}$ is defined by complementing the $V_t^{p,q}$ entry in $\mathcal{X}_t$.

A joint probability function of two random variables, $\mathcal{W}_t^P$, is defined over the discretized $\mathbb{R}^+$, and $\mathcal{X}_t$ defined over $S(P,t)$, e.g., $f_{(\mathcal{W}_t^P,\mathcal{X}_t)}(w, (V_t^{p_1,q_1}, V_t^{p_2,q_2}, \ldots, V_t^{p_n,q_n})) = \mathbb{P}(\mathcal{W}_t^P = w, \mathcal{X}_t = (V_t^{p_1,q_1}, V_t^{p_2,q_2}, \ldots, V_t^{p_n,q_n}))$. The original pmf for $\mathcal{W}_t^P$ can be reconstructed from the joint pmfs:

$$f_{\mathcal{W}_t^P}(\cdot) = \sum_{x \in S(P,t)} f_{(\mathcal{W}_t^P,\mathcal{X}_t)}(\cdot, \mathcal{X}_t = x). \tag{4.3}$$

We are interested in computing the backlog $\mathcal{W}_t^{P_i}$ for an instance $M_{i,j}$ of m_i (actual message, not characteristic), where t is any instant after its queuing time ($t \geq \mathcal{Q}_{i,j}$) as a function of the interference set of $M_{i,j}$ or $I(P_i, t)$.

4.4.2 Stochastic Analysis of the Approximate System

In order to compute the pmf of the message response-time in the approximate system, first the stationary distribution of the backlog at the beginning of the hyperperiod is computed; then, the pmf of the backlog at the queuing time of each

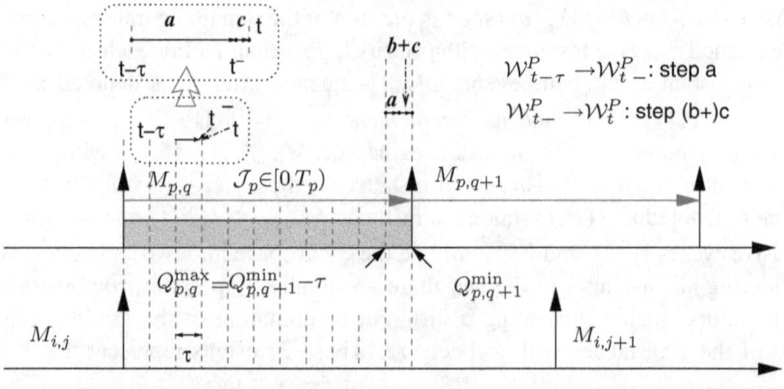

Fig. 4.4 Updating the backlog in the discrete-time model

message instance, and finally, the pmf of the response-time of each *message instance* within the hyperperiod are computed. The message response-time pmf is obtained by averaging the response-time pmfs of all the instances in the hyperperiod.

4.4.2.1 Stationary Backlog Within the Hyperperiod

The example in Fig. 4.4 illustrates the method used to update the backlog at priority level P_i for m_i. The characteristic message instances, which represent the load from remote ECUs (such as $M_{p,q}$ in Fig. 4.4), are queued with random jitter in $[0, T_p)$. At any time t, the backlog is computed knowing its value at the previous tick $t - \tau$ and going through an intermediate step, an instant t^- arbitrarily close to t (this step is enlarged in the bubble on the top-left side of Fig. 4.4). Starting from the backlog at time $t - \tau$, first the backlog is updated by shrinking (step **a** in Fig. 4.4) to time t^-, right before the possible queuing time of message instance $M_{p,q} \in I(P, t)$. By shrinking, the time is advanced, accumulating in the origin (backlog equal to zero) all the probabilities of non-positive backlogs. The probability that a message instance still to be activated is queued at t, $\mathbb{P}(\mathcal{Q}_{p,q} = t | \mathcal{Q}_{p,q} \geq t)$ can be easily computed as $1/n_t$ where n_t is the number of ticks from t to the latest possible queuing time $Q^{max}_{p,q}$ for $M_{p,q}$.

Finally, the rules are provided to compute the backlog from t^- to t in two possible scenarios, as a result of the possible (probabilistic) queuing of characteristic messages from remote nodes at t. One possible scenario (step **c** in Fig. 4.4) is when the set of message instances that can contribute to the backlog remains the same as that of $t - \tau$. The other case is when at time $t = Q^{min}_{p,q+1} = Q^{max}_{p,q} + \tau$, there is at least one new instance $M_{p,q+1}$ that must be considered in place of $M_{p,q}$. In this case, an additional step, labeled as **b** in Fig. 4.4, must be performed after the shrinking and before considering the backlog update. The following subsections describe the procedure for updating the backlog in these two cases: possible queuing of a message, and a change in the set of message instances.

Updating the Backlog for a Possible Queuing of a Message Instance

To find the probability of $\mathcal{W}_t^P = \mathcal{W}_{t^-}^P + e = w$ where e is either 0 or a possible contribution to the backlog by a message instance $M_{p,q}$, we consider the joint pmfs of $\mathcal{W}_t^P$ and the queuing patterns $\overline{V}_t^{p,q}$ and $V_t^{p,q}$, that is, $f_{(\mathcal{W}_t^P,\overline{V}_t^{p,q})}$ and $f_{(\mathcal{W}_t^P,V_t^{p,q})}$:

- Scenario (1): $\mathcal{W}_t^P = w$ and $\overline{V}_t^{p,q}$: $(\mathcal{Q}_{p,q} \leq t)$, that is, the backlog at time t at priority higher than P is w *and* the new message instance $M_{p,q}$ has been queued at time prior to or at t. Then, either $M_{p,q}$ contributed to the backlog at any time prior or equal to t^-, and in this case $e = 0$, that is, $M_{p,q}$'s transmission time $\mathcal{E}_p$ has been already included in $\mathcal{W}_{t^-}^P$, or $M_{p,q}$ is queued at t, and the sum of its transmission time $\mathcal{E}_p$ and the backlog at t^- must be equal to w.
- Scenario (2): $\mathcal{W}_t^P = w$ and $V_t^{p,q}$: $(\mathcal{Q}_{p,q} > t)$, that is, the backlog at time t at priority higher than P is w *and* the new message instance $M_{p,q}$ is queued after t and does not contribute to $\mathcal{W}_t^P$. Therefore, $\mathcal{E}_p$ is not part of the backlog at time t, and $\mathcal{W}_t^P = \mathcal{W}_{t^-}^P = w$.

The following theorem formulates the backlog update process.

Theorem 4.4.2. *Consider a characteristic message instance $M_{p,q}$, possibly queued at time t, with priority higher than P. $\mathcal{W}_{t^-}^P$ is the P-level backlog immediately before t. The probability that the backlog at time t equals a given value w is computed with respect to the mutually exclusive events $\overline{V}_t^{p,q}$ and $V_t^{p,q}$.*

$$\begin{aligned}
&\mathbb{P}\left(\mathcal{W}_t^P = w, \overline{V}_t^{p,q}\right)\\
&= \mathbb{P}\left(\mathcal{W}_{t^-}^P = w, \overline{V}_{t^-}^{p,q}\right) + \textstyle\sum_{e=0}^{w} \mathbb{P}\left(\mathcal{W}_{t^-}^P = w - e, V_{t^-}^{p,q}\right)\\
&\times \mathbb{P}\left(\mathcal{Q}_{p,q} = t | \mathcal{Q}_{p,q} \geq t\right) \times \mathbb{P}\left(\mathcal{E}_p = e\right)\\
&\mathbb{P}\left(\mathcal{W}_t^P = w, V_t^{p,q}\right)\\
&= \mathbb{P}\left(\mathcal{W}_{t^-}^P = w, V_{t^-}^{p,q}\right) \times \mathbb{P}\left(\mathcal{Q}_{p,q} > t | \mathcal{Q}_{p,q} \geq t\right)\\
&= \mathbb{P}\left(\mathcal{W}_{t^-}^P = w, V_{t^-}^{p,q}\right) \times \left(1 - \mathbb{P}\left(\mathcal{Q}_{p,q} = t | \mathcal{Q}_{p,q} \geq t\right)\right)
\end{aligned}$$

The computation of the joint pmfs of the P-level backlog with respect to the two events $\overline{V}_t^{p,q}$ and $V_t^{p,q}$ can be performed as follows ($f_{\mathcal{V}_1} \otimes f_{\mathcal{V}_2}$ denotes the convolution of $f_{\mathcal{V}_1}$ and $f_{\mathcal{V}_2}$):

$$\begin{aligned}
f_{(\mathcal{W}_t^P,\overline{V}_t^{p,q})} &= f_{(\mathcal{W}_{t^-}^P,\overline{V}_{t^-}^{p,q})} + \mathbb{P}\left(\mathcal{Q}_{p,q} = t | \mathcal{Q}_{p,q} \geq t\right) \times \left(f_{(\mathcal{W}_{t^-}^P,V_{t^-}^{p,q})} \otimes f_{\mathcal{E}_p}\right)\\
f_{(\mathcal{W}_t^P,V_t^{p,q})} &= \left(1 - \mathbb{P}\left(\mathcal{Q}_{p,q} = t | \mathcal{Q}_{p,q} \geq t\right)\right) \times f_{(\mathcal{W}_{t^-}^P,V_{t^-}^{p,q})}.
\end{aligned}$$

Proof. The proof can be found in [63]. □

In general, there may be more than one message instance queued at time t, therefore it is necessary to consider the joint pmfs of the backlog with all the queuing patterns $\mathcal{X}_t \in S(P,t)$. Using 4.4 we compute the joint pmf of the P-level backlog at t together with each pattern $\mathcal{X}_t$, after the possible queuing of a message instance $M_{p,q} \in I(P,t)$. In case $\mathcal{X}_t$ is a negative pattern of $V_t^{p,q}$, the pmf

Algorithm 2 Cross-convolution: calculate P-level backlog after the possible queuing of $M_{p,q}$ at time t

for each **negative** pattern $\overline{\mathcal{X}}_t \in S(P,t)$ of $V_t^{p,q}$ **do**
  $\mathcal{X}_{t(p,q)}$ = complementary pattern of $\overline{\mathcal{X}}_t$ wrt $M_{p,q}$
  $f_{(\mathcal{W}_t^P,\overline{\mathcal{X}}_t)} = f_{(\mathcal{W}_{t^-}^P,\overline{\mathcal{X}}_{t^-})} + \mathbb{P}\left(\mathcal{Q}_{p,q} = t | \mathcal{Q}_{p,q} \geq t\right) \times \left(f_{(\mathcal{W}_{t^-}^P,\mathcal{X}_{t^-(p,q)})} \otimes f_{\mathcal{E}_p}\right)$
end for
for each **positive** pattern $\mathcal{X}_t \in S(P,t)$ of $V_t^{p,q}$ **do**
  $f_{(\mathcal{W}_t^P,\mathcal{X}_t)} = \left(1 - \mathbb{P}\left(\mathcal{Q}_{p,q} = t | \mathcal{Q}_{p,q} \geq t\right)\right) \times f_{(\mathcal{W}_{t^-}^P,\mathcal{X}_{t^-})}$
end for

$f_{(\mathcal{W}_t^P,\mathcal{X}_t)}$ can be obtained by computing the convolution of the joint pmf defined with respect to the complementary queuing pattern $\overline{\mathcal{X}}_{t^-(p,q)}$, $f_{(\mathcal{W}_{t^-}^P,\overline{\mathcal{X}}_{t^-(p,q)})}$, with the transmission time pmf $f_{\mathcal{E}_p}$ of $M_{p,q}$, multiplied by $\mathbb{P}\left(\mathcal{Q}_{p,q} = t | \mathcal{Q}_{p,q} \geq t\right)$, and adding the result to the joint pmf $f_{(\mathcal{W}_{t^-}^P,\mathcal{X}_{t^-})}$ (by extension of (4.4)). In case $\mathcal{X}_t$ is a positive pattern of $V_t^{p,q}$, $f_{(\mathcal{W}_t^P,\mathcal{X}_t)}$ is obtained by multiplying $f_{(\mathcal{W}_{t^-}^P,\mathcal{X}_{t^-})}$ with $1 - \mathbb{P}\left(\mathcal{Q}_{p,q} = t | \mathcal{Q}_{p,q} \geq t\right)$ (by extension of (4.4)). This operation is called *cross-convolution* since the update of the joint pmf $f_{(\mathcal{W}_t^P,\mathcal{X}_t)}$ for a negative pattern $\mathcal{X}_t$ of $V_t^{p,q}$ is performed by adding a term resulting from the convolution of the transmission time pmf of $M_{p,q}$ with the joint pmf of the backlog and the complementary queuing pattern $\overline{\mathcal{X}}_{t^-(p,q)}$ with respect to $M_{p,q}$, as described in Algorithm 2.

Queuing Pattern Update

A special case occurs at the time instants when the sets of P-level message instances change. A characteristic message instance $M_{p,q}$ is replaced by $M_{p,q+1}$ at time $t = \mathcal{Q}_{p,q+1}^{\min} = \mathcal{Q}_{p,q}^{\max} + \tau$. The entries $V_{t^-}^{p,q}$ and $\overline{V}_{t^-}^{p,q}$, in the queuing pattern vector $\mathcal{X}_t$, must be replaced by two new queuing events $V_{t^-}^{p,q+1}$ and $\overline{V}_{t^-}^{p,q+1}$. The joint pmfs $f_{(\mathcal{W}_{t^-}^P,\mathcal{X}_{t^-})}$ for the new patterns can be defined starting from the consideration that $\mathbb{P}\left(\overline{V}_{t^-}^{p,q+1}\right) = \mathbb{P}\left(V_{t^-}^{p,q}\right) = 0$ and $\mathbb{P}\left(V_{t^-}^{p,q+1}\right) = \mathbb{P}\left(\overline{V}_{t^-}^{p,q}\right) = 1$. Hence, the pmf $f_{(\mathcal{W}_{t^-}^P,\mathcal{X}'_{t^-})}$ for each queuing pattern $\mathcal{X}'_{t^-}$ that includes $\overline{V}_{t^-}^{p,q+1}$, that is, $\mathcal{X}'_{t^-} = \{\ldots, \overline{V}_{t^-}^{p,q+1}, \ldots\}$ can be defined as $\mathbb{P}\left(\mathcal{W}_{t^-}^P = w, \mathcal{X}'_{t^-}\right) = 0$, $\forall w$. The values of $f_{(\mathcal{W}_{t^-}^P,\mathcal{X}''_{t^-})}$ for the other patterns that include the complement $V_{t^-}^{p,q+1}$ can be obtained by setting $f_{(\mathcal{W}_{t^-}^P,\mathcal{X}''_{t^-})} = f_{(\mathcal{W}_{t^-}^P,\mathcal{X}^*_{t^-})}$ where all the elements of $\mathcal{X}^*_{t^-}$ match the corresponding elements of $\mathcal{X}''_{t^-}$, except that $\overline{V}_{t^-}^{p,q}$ is replaced in $\mathcal{X}''_{t^-}$ by $V_{t^-}^{p,q+1}$.

Thus, given the stationary P-level backlog pmf at the beginning of the hyperperiod, the stationary P-level backlog pmf at any time in the hyperperiod can be computed by iteratively using the operations of shrinking, queuing pattern update (possibly) and cross-convolution.

4.4.2.2 Initial Blocking Time

Besides the delay from higher priority messages, a message is also subject to an initial blocking delay from lower priority messages due to the impossibility of preempting message transmissions. Given an instance $M_{i,j}$, we consider the set of all lower priority messages, denoted as $lp(P_i) = \{m_k | P_k > P_i\}$, which cause a blocking delay $\mathcal{B}_{i,j}$ to $M_{i,j}$ if they are transmitted at the time $M_{i,j}$ is queued. A blocking time of length $b > 0$ is caused if a message in $lp(P_i)$ has transmission time $\mathcal{E}_k > b$ and its transmission starts exactly $\mathcal{E}_k - b$ units before the queuing instant of $M_{i,j}$. If we assume that the transmissions of instances in $lp(P_i)$ are evenly distributed in the hyperperiod, then in the discrete-time system, the probability that one such instance transmission starts at exactly the right time tick in the hyperperiod is $\frac{\tau}{T_k}$. The probability that $M_{i,j}$ waits for a time $b > 0$ is obtained by adding up all the contributions from the messages $m_k \in lp(P_i)$

$$\mathbb{P}(\mathcal{B}_{i,j} = b) = \sum_{m_k \in lp(P_i)} \frac{\mathbb{P}(\mathcal{E}_k > b)}{T_k/\tau}. \tag{4.4}$$

Given these probabilities, the pmf $f_{\mathcal{B}_{i,j}}$ of $\mathcal{B}_{i,j}$ is easily computed and the blocking time is added (by convolution) to the backlog at the message queuing time. *Additional errors are introduced in this step, given that transmissions of lower priority messages are actually not uniformly distributed in the hyperperiod.* The initial blocking time is not included in the approximate system and does not affect its stability.

4.4.2.3 Message Response-Time Calculation

The computation of the response-time of $M_{i,j}$ requires, as a first step, the evaluation of its backlog at its queuing time. The other local messages are queued periodically at known instants, and the remote messages are represented by their P_i-level characteristic messages, queued periodically with random jitter. The backlog $\mathcal{W}^{P_i}$ is computed by applying the methods previously defined, and then the backlog at the queuing time of $M_{i,j}$ is updated to account for a possible initial blocking.

The earliest possible start time for $M_{i,j}$ is its queuing time $t = \mathcal{Q}_{i,j}$. Starting from t and, considering the following times $t_k = t + k\tau$ ($k \in N^+$), we further decompose $f_{\left(\mathcal{W}_{t_k}^{P_i}, \mathcal{X}_{t_k}\right)}$ into the joint pmfs $f_{\left(\mathcal{W}_{t_k}^{P_i}, \mathcal{S}_{i,j} < t_k, \mathcal{X}_{t_k}\right)}$ and $f_{\left(\mathcal{W}_{t_k}^{P_i}, \mathcal{S}_{i,j} \geq t_k, \mathcal{X}_{t_k}\right)}$, and the latter in $f_{\left(\mathcal{W}_{t_k}^{P_i}, \mathcal{S}_{i,j} = t_k, \mathcal{X}_{t_k}\right)}$ and $f_{\left(\mathcal{W}_{t_k}^{P_i}, \mathcal{S}_{i,j} > t_k, \mathcal{X}_{t_k}\right)}$.

At time t ($k = 0$), and for each queuing pattern $\mathcal{X}_t \in S(P_i, t)$, clearly $f_{\left(\mathcal{W}_t^{P_i}, \mathcal{S}_{i,j} \geq t, \mathcal{X}_t\right)} = f_{\left(\mathcal{W}_t^{P_i}, \mathcal{X}_t\right)}$ (the start time is no earlier than the queuing time). The probability that $M_{i,j}$ starts transmission immediately, i.e., $\mathbb{P}\left(\mathcal{S}_{i,j} = \mathcal{Q}_{i,j} = t\right)$ equals the probability that the backlog of $M_{i,j}$ at t is zero.

$$\mathbb{P}(\mathcal{S}_{i,j} = t) = \mathbb{P}\left(\mathcal{W}_t^{P_i} = 0\right) = \sum_{\mathcal{X}_t} \mathbb{P}\left(\mathcal{W}_t^{P_i} = 0, \mathcal{S}_{i,j} \geq t, \mathcal{X}_t\right)$$

Given that $\mathcal{W}_t^{P_i} = 0$ implies $\mathcal{S}_{i,j} = t$, and $\mathcal{W}_t^{P_i} = w$ with $w > 0$ implies $\mathcal{S}_{i,j} > t$, we have

$$\mathbb{P}\left(\mathcal{W}_t^{P_i} = 0, \mathcal{S}_{i,j} > t\right) = 0,$$
$$\mathbb{P}\left(\mathcal{W}_t^{P_i} = w, \mathcal{S}_{i,j} > t\right) = \mathbb{P}\left(\mathcal{W}_t^{P_i} = w, \mathcal{S}_{i,j} \geq t\right), \forall w > 0$$

which means that $f_{\left(\mathcal{W}_t^{P_i}, \mathcal{S}_{i,j} > t, \mathcal{X}_t\right)}\left(\mathcal{W}_t^{P_i} = 0\right) = 0$, and, for all $w > 0$, it is $f_{\left(\mathcal{W}_t^{P_i}, \mathcal{S}_{i,j} > t, \mathcal{X}_t\right)}\left(\mathcal{W}_t^{P_i} = w\right) = f_{\left(\mathcal{W}_t^{P_i}, \mathcal{S}_{i,j} \geq t, \mathcal{X}_t\right)}\left(\mathcal{W}_t^{P_i} = w\right)$ for each $\mathcal{X}_t \in S(M_{i,j}, t)$. Also, since $\mathcal{S}_{i,j} > t$ implies $\mathcal{W}_t^{P_i} > 0$, it is

$$\sum_w \mathbb{P}\left(\mathcal{W}_t^{P_i} = w, \mathcal{S}_{i,j} > t\right) = 1 - \mathbb{P}(\mathcal{S}_{i,j} = t). \tag{4.5}$$

The probability that $M_{i,j}$ starts transmission at the next time instants $t_k = t + k\tau$ can be computed by iteratively applying the following steps. Starting from $f_{\left(\mathcal{W}_t^{P_i}, \mathcal{S}_{i,j} > t, \mathcal{X}_t\right)}$, the compound pmfs of the backlog is updated to the next time point $t_1 = t + \tau$ by applying the methods defined in Sect. 4.4.2.1. In the discrete system, $\mathcal{S}_{i,j} > t$ is the same as $\mathcal{S}_{i,j} \geq t_1$, hence, $f_{\left(\mathcal{W}_{t_1}^{P_i}, \mathcal{S}_{i,j} \geq t_1\right)} = f_{\left(\mathcal{W}_{t_1}^{P_i}, \mathcal{S}_{i,j} > t\right)}$. Since the operations of cross-convolution, shrinking and queuing pattern update do not change the total probability of the backlog, thus

$$\sum_w \mathbb{P}\left(\mathcal{W}_{t_1}^{P_i} = w, \mathcal{S}_{i,j} \geq t_1\right) = 1 - \mathbb{P}\left(\mathcal{S}_{i,j} = t\right)$$

The probability that $M_{i,j}$ starts transmission at time t_1 equals the probability that its backlog is zero at t_1

$$\mathbb{P}\left(\mathcal{S}_{i,j} = t_1\right) = \mathbb{P}\left(\mathcal{W}_{t_1}^{P_i} = 0\right) = \sum_{\mathcal{X}_{t_1}} \mathbb{P}\left(\mathcal{W}_{t_1}^{P_i} = 0, \mathcal{S}_{i,j} \geq t_1, \mathcal{X}_{t_1}\right).$$

Similar to time instant t, we compute the compound pmf for the case $\mathcal{S}_{i,j} > t_1$ as $f_{\left(\mathcal{W}_{t_1}^{P_i}, \mathcal{S}_{i,j} > t_1, \mathcal{X}_{t_1}\right)}(0) = 0$, and $\forall w > 0$, $f_{\left(\mathcal{W}_{t_1}^{P_i}, \mathcal{S}_{i,j} > t_1, \mathcal{X}_{t_1}\right)}(w) = f_{\left(\mathcal{W}_{t_1}^{P_i}, \mathcal{S}_{i,j} \geq t_1, \mathcal{X}_{t_1}\right)}(w)$. Iterating (4.5), the total probability of $f_{\left(\mathcal{W}_{t_1}^{P_i}, \mathcal{S}_{i,j} > t_1\right)}$ is

$$\sum_w \mathbb{P}\left(\mathcal{W}_{t_1}^{P_i} = w, \mathcal{S}_{i,j} > t_1\right) = 1 - \sum_{t_k \leq t_1} \mathbb{P}\left(\mathcal{S}_{i,j} = t_k\right).$$

Algorithm 3 Calculate the start time pmf of $M_{i,j}$

$t = Q_{i,j}$
for each queuing pattern $\mathcal{X}_t \in S(M_{i,j}, t)$ **do**
  $f_{(\mathcal{W}_t^{P_i}, \mathcal{S}_{i,j} \ge t, \mathcal{X}_t)} = f_{(\mathcal{W}_{t^-}^{P_i}, \mathcal{X}_t)}$
end for
$\forall k \ge Q_{i,j}, \mathbb{P}(\mathcal{S}_{i,j} = k) = 0$
while $\sum_{k=Q_{i,j}}^{t} \mathbb{P}\left(\mathcal{S}_{i,j} = k\right) < 1$ **do**
  for each queuing pattern $\mathcal{X}_t \in S(M_{i,j}, t)$ **do**
    $\mathbb{P}\left(\mathcal{S}_{i,j} = t\right) += \mathbb{P}\left(\mathcal{W}_t^{P_i} = 0, \mathcal{S}_{i,j} \ge t, \mathcal{X}_t\right)$
    $\forall w > 0, \mathbb{P}\left(\mathcal{W}_t^{P_i} = w, \mathcal{S}_{i,j} > t, \mathcal{X}_t\right) = \mathbb{P}\left(\mathcal{W}_t^{P_i} = w, \mathcal{S}_{i,j} \ge t, \mathcal{X}_t\right)$
  end for
  Update $f_{(\mathcal{W}_t^{P_i}, \mathcal{S}_{i,j} > t)}$ to get $f_{(\mathcal{W}_{t+\tau}^{P_i}, \mathcal{S}_{i,j} > t)}$, i.e., $f_{(\mathcal{W}_{t+\tau}^{P_i}, \mathcal{S}_{i,j} \ge t+\tau)}$
  $t = t + \tau$
end while

The above two steps for $t_k = t + k\tau$ are iteratively applied until we get to time t_n when the compound backlog is null.

$$\forall \mathcal{X}_n \in S(M_{i,j}, t_n),\ \forall w > 0,\ \mathbb{P}\left(\mathcal{W}_{t_n}^{P_i} = w, \mathcal{S}_{i,j} \ge t_n, \mathcal{X}_n\right) = 0.$$

This happens when, at time t_n, the total probability of the start time pmf accumulates to 1 (as outlined in Algorithm 3).

$$\sum_{k \le n} \mathbb{P}(\mathcal{S}_{i,j} = t_k) = 1.$$

Once the pmf of the time $\mathcal{S}_{i,j}$ at which $M_{i,j}$ starts its transmission is known, the pmf of its finish time is obtained by simply adding its transmission time $\mathcal{E}_i$, i.e., performing a convolution on $f_{\mathcal{S}_{i,j}}$ and $f_{\mathcal{E}_i}$. The response-time is finally computed by a further left shift of $Q_{i,j}$ time units.

$$\begin{aligned} f_{\mathcal{F}_{i,j}} &= f_{\mathcal{S}_{i,j}} \otimes f_{\mathcal{E}_i} \\ \mathbb{P}\left(\mathcal{R}_{i,j} = t - Q_{i,j}\right) &= \mathbb{P}\left(\mathcal{F}_{i,j} = t\right), \forall t \end{aligned} \tag{4.6}$$

The response-time pmf $f_{\mathcal{R}_i}$ of m_i is the arithmetic average of the pmfs $f_{\mathcal{R}_{i,j}}$ of its instances $M_{i,j}$ inside the hyperperiod.

4.4.2.4 Algorithm Complexity

This section analyzes the worst-case complexity of the algorithm to compute the backlog of level i for message instance $M_{i,j}$ on node ECU$_n$. The algorithm refers to the computations in one hyperperiod and requires at each step the evaluation of cross convolution and shrinking. Let n be the number of nodes in the system. Within each hyperperiod, there are $(n-1)H/\tau+\sum_{k\in hp(i,n)} H/T_k$ possible queuing events of higher priority message instances (the number is $O((n-1)H/\tau)$ since $T_k \gg \tau$). A backlog update requires updating 2^{n-1} compound backlogs, 2^{n-2} of which along with positive queuing patterns. Each of these compound backlogs requires considering, at each step, each possible queuing of a message instances by performing a cross-convolution. The complexity of convolution depends on the algorithm used and on the worst-case length of the distributions. In the analysis, the maximum execution time of each characteristic message and the compound backlogs are less than $R_i^{\max}$. To perform a convolution on two vectors f and g, with size n_f and n_g respectively, the number of operations required is $O(n_f n_g)$.[3] Thus the complexity of calculation within one hyperperiod is $O(2^{n-2} \times (R_i^{\max}/\tau)^2 \times (n-1)H/\tau) = O\left(\frac{n2^n(R_i^{\max})^2 H}{\tau^3}\right)$. This is a worst-case bound, and the algorithm is typically much faster in the average case.

4.5 Analysis on an Example Automotive System

As an example, the results on one of the CAN buses of an experimental vehicle is presented in this section. The test set consists of 6 ECUs and 69 messages as shown in Table 4.1. T_i is in milliseconds, the message size is in bytes, and the transmission time distribution is calculated using the independent bit-stuffing model as in [51]. All message offsets are zero. The bus rate is 500 kbps, and the maximum utilization of the system is 60.25%.

The quality of the stochastic analysis is demonstrated by comparing its results with simulations. It takes less than 2 h to analyze all the messages, with a maximum of 7 min each with $\tau = 10\mu$ s. In simulation, the relative phases among nodes are sampled with granularity of 50μ s, leading to 3.2×10^{18} possible combinations. This space cannot be explored exhaustively, hence, 10^8 relative phases are randomly chosen, for 10 h of total running time. The stationary distribution of backlogs is obtained by iterative approximation, that is, stopping when the pmfs of backlogs are close enough at the end of two consecutive hyperperiods. The stationary distribution of the messages response-times are calculated in the next hyperperiod.

The results for four representative messages, two with relatively high priority, and the other two with low priority, are presented, and the cumulative distribution functions (cdfs) of the stochastic analysis are compared with simulation, along with the values from worst-case analysis [22].

For the first representative message m_5, the worst-case response-time is shorter than half of the gcd of the periods (2.5 ms). Hence, the analysis is very accurate as the actual backlog does not include any multiple message instances from any

[3] The operation of convolution can be performed faster using Fast Fourier Transform algorithm.

Table 4.1 An automotive CAN system with 6 ECUs and 69 messages

m_i	Υ_i^{src}	T_i	Size	P_i	m_i	Υ_i^{src}	T_i	Size	P_i	m_i	Υ_i^{src}	T_i	Size	P_i
m_1	ECU_2	10	8	1	m_{24}	ECU_3	25	6	24	m_{47}	ECU_2	50	4	47
m_2	ECU_2	10	8	2	m_{25}	ECU_3	25	7	25	m_{48}	ECU_4	100	8	48
m_3	ECU_3	5	4	3	m_{26}	ECU_2	20	8	26	m_{49}	ECU_3	100	8	49
m_4	ECU_3	10	7	4	m_{27}	ECU_5	25	8	27	m_{50}	ECU_1	100	8	50
m_5	ECU_1	10	4	5	m_{28}	ECU_2	20	8	28	m_{51}	ECU_3	100	1	51
m_6	ECU_6	10	8	6	m_{29}	ECU_1	25	3	29	m_{52}	ECU_3	100	8	52
m_7	ECU_1	100	8	7	m_{30}	ECU_2	10	5	30	m_{53}	ECU_1	100	4	53
m_8	ECU_1	100	2	8	m_{31}	ECU_2	20	8	31	m_{54}	ECU_3	100	1	54
m_9	ECU_1	100	3	9	m_{32}	ECU_4	10	8	32	m_{55}	ECU_3	100	1	55
m_{10}	ECU_5	25	8	10	m_{33}	ECU_3	10	8	33	m_{56}	ECU_5	100	8	56
m_{11}	ECU_1	100	2	11	m_{34}	ECU_3	10	8	34	m_{57}	ECU_3	100	7	57
m_{12}	ECU_1	20	4	12	m_{35}	ECU_1	25	7	35	m_{58}	ECU_3	100	1	58
m_{13}	ECU_2	100	4	13	m_{36}	ECU_6	25	6	36	m_{59}	ECU_3	100	1	59
m_{14}	ECU_1	100	4	14	m_{37}	ECU_4	50	8	37	m_{60}	ECU_4	100	3	60
m_{15}	ECU_5	100	8	15	m_{38}	ECU_4	50	8	38	m_{61}	ECU_1	100	8	61
m_{16}	ECU_1	10	6	16	m_{39}	ECU_5	50	8	39	m_{62}	ECU_1	100	1	62
m_{17}	ECU_2	100	7	17	m_{40}	ECU_3	50	5	40	m_{63}	ECU_3	100	4	63
m_{18}	ECU_2	100	8	18	m_{41}	ECU_4	50	8	41	m_{64}	ECU_1	100	8	64
m_{19}	ECU_2	50	7	19	m_{42}	ECU_6	50	8	42	m_{65}	ECU_1	100	1	65
m_{20}	ECU_3	10	8	20	m_{43}	ECU_2	50	8	43	m_{66}	ECU_1	100	1	66
m_{21}	ECU_6	10	8	21	m_{44}	ECU_5	100	8	44	m_{67}	ECU_2	50	8	67
m_{22}	ECU_6	25	8	22	m_{45}	ECU_2	25	2	45	m_{68}	ECU_1	100	1	68
m_{23}	ECU_3	25	6	23	m_{46}	ECU_2	50	4	46	m_{69}	ECU_4	100	8	69

remote characterization message. As expected (top figure in Fig. 4.5) the cumulative distribution function (cdf) obtained from the analysis and the simulation match very well. In the figure, the analysis of the response-time of the high priority message 1 has been tentatively performed without the additional contribution to the backlog caused by the blocking factor. The result is the dashed line, which is quite optimistic with respect to the actual values and shows the accuracy in the modeling of this term.

For the second representative m_{25}, the worst-case response-time is slightly larger than the gcd of the periods (5 ms), but the analysis is quite accurate. The results start to deviate from the simulation when the worst-case response-time grows longer. Message m_{43} and m_{63} have a worst-case response-time significantly larger than the gcd of the periods. However, the distribution obtained from the analysis is still a good match to that of the simulation (Fig. 4.6). For low priority messages, the probabilities computed by the stochastic analysis for long response-times are larger than those computed by the simulation as the queuing times of the characteristic messages are a pessimistic approximation with respect to interferences (consecutive instances of a characteristic message can be separated by less than its period). Furthermore, as shown in both figures, the cumulative probability approaches 100% probability at response-time values significantly smaller than the worst-case value.

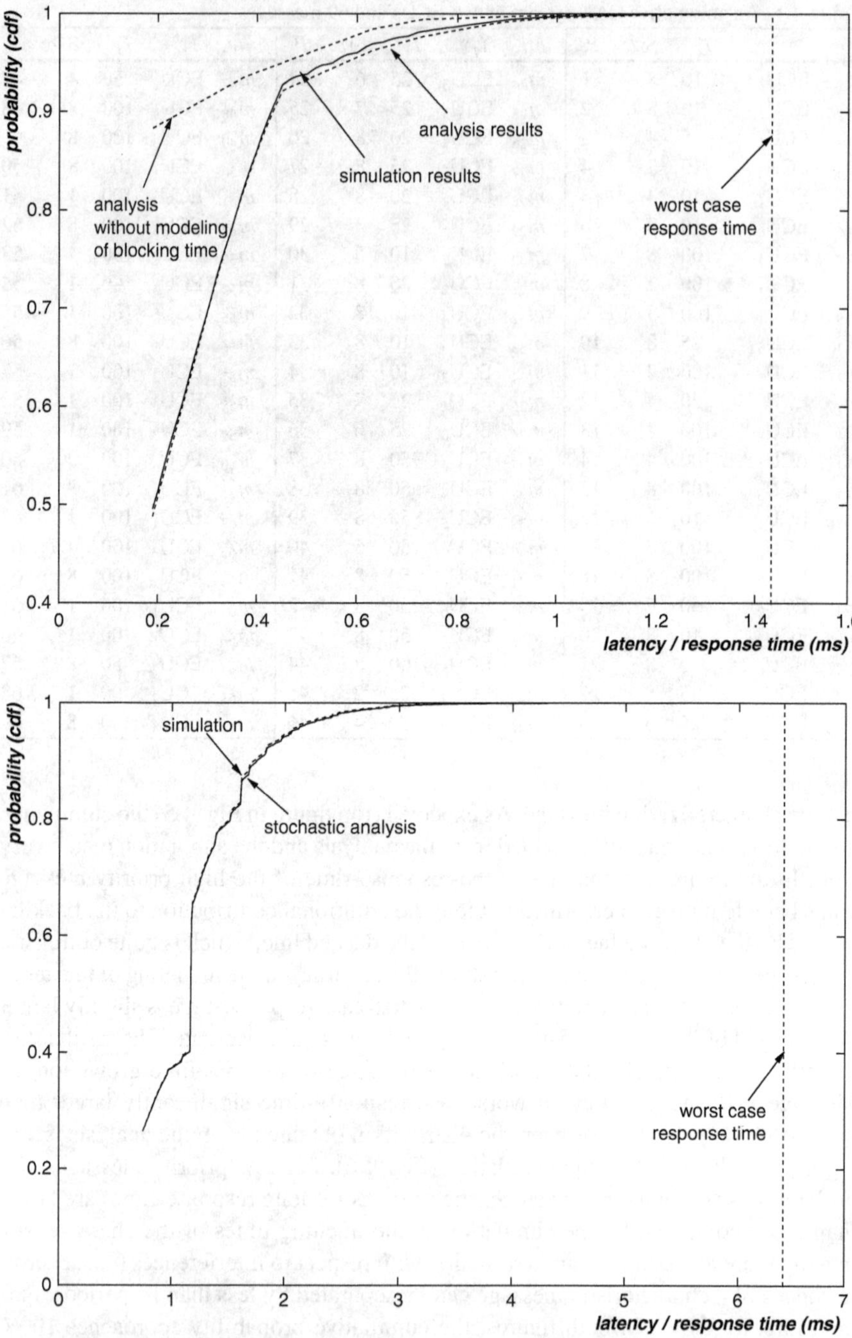

Fig. 4.5 The response-time cdfs of high priority messages m_5 and m_{25} in Table 4.1

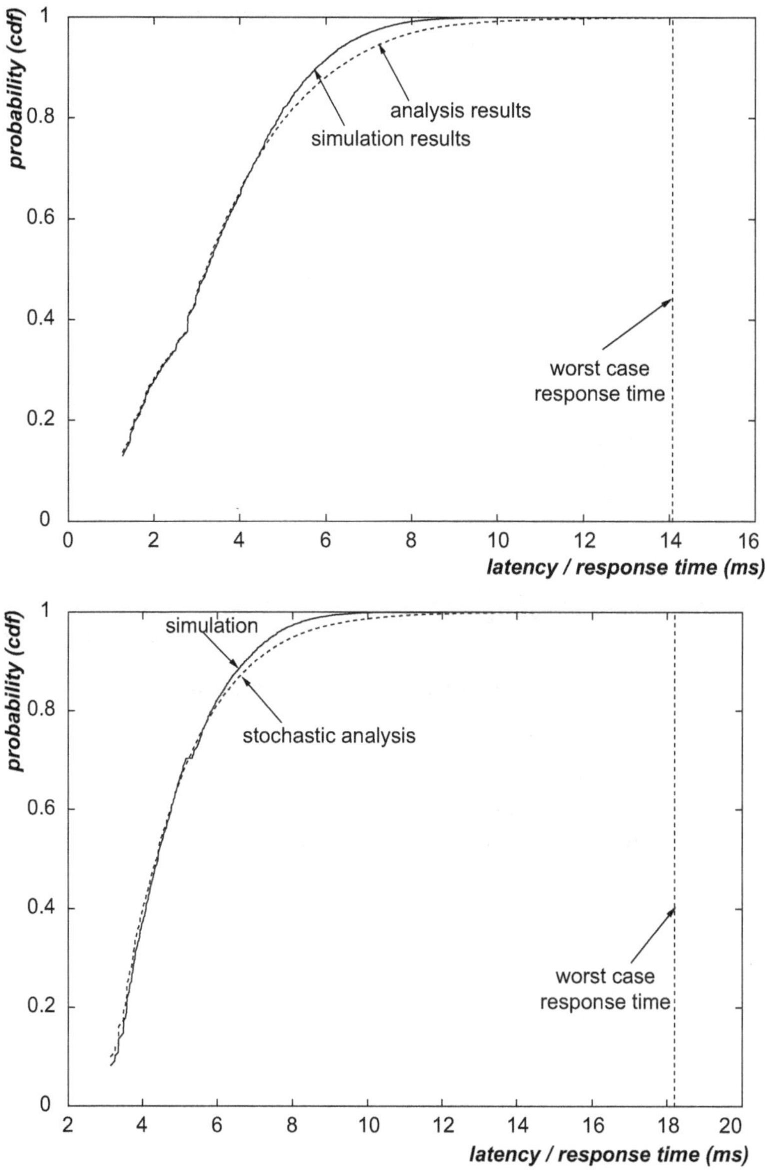

Fig. 4.6 The response-time cdfs of low priority messages m_{43} and m_{63} in Table 4.1

Finally, Fig. 4.7 shows the results of the message response-time analysis of a complex message trace extracted from a product vehicle (not simulation data). The measured response-time cdf of a low priority message (CAN id $> 0 \times 500$) is compared with the response-time estimated by the stochastic analysis. The

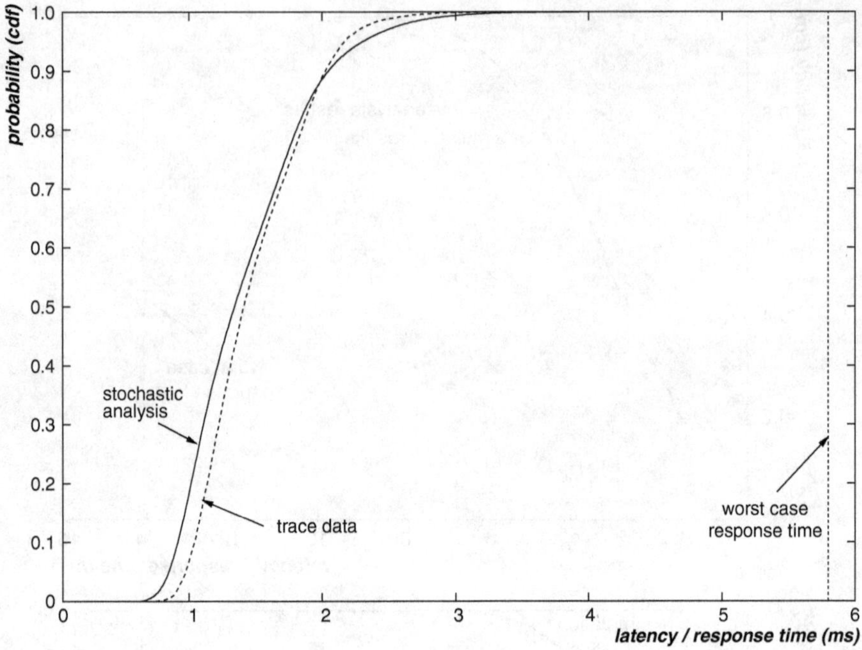

Fig. 4.7 The response-time cdf of a low priority message in a bus trace

analysis is accurate despite the large worst-case message response-time and the non-idealities of the actual implementation (possible priority inversions at the network adapter and significant copy times between the message queues and the TxObjects).

Chapter 5
Statistical Analysis

After worst-case analysis and stochastic analysis, another alternative for the timing evaluation of CAN systems is the use of statistical analysis, possibly in conjunction with simulation, or leveraging data from traces and simulations. The statistical analysis of message response-times can be applied in at least two contexts. The first consists in the analysis of an existing system, using trace data or simulation data to compute the probability of latency values for selected messages. The second is the use of the regression formulas presented in this chapter to predict the probability of message latencies in a CAN system for which only limited information is available, for example, when a hypothetical architecture option is analyzed. In the latter case, the analysis is for a CAN bus as part of a larger system, in which end-to-end response-times need to be estimated to provide a measure of the quality of the architecture solution (see the discussion at the end of Chap. 3). A special domain of interest for this second case is the early evaluation of automotive architectures.

5.1 Introduction

An automotive electrical/electronic (E/E) is defined and used over several years. The basic architecture configuration needs to be defined and frozen years in advance of production, when the communication and computation load is partly unknown. Later, during the architecture lifespan, functions are placed on ECUs and communication is scheduled on the bus. Hence, in the early stages, performance (e.g., latency and utilization) of architectures are measured against loads that have been estimated by from past trends or early indications of future need. The statistical analysis techniques presented in this chapter can be used for the early evaluation and selection of E/E architecture for next-generation automotive controls, where the application performance depends on the end-to-end latencies of active-safety functions. The results show that it is possible to make tactical decisions (e.g., message priority assignment) in the early phases of the development process when

M. Di Natale et al., *Understanding and Using the Controller Area Network Communication Protocol: Theory and Practice*, DOI 10.1007/978-1-4614-0314-2_5,

incomplete and estimated information (e.g., the current/estimated bus utilization or the maximum queue length on the bus adapter side) are available.

In this chapter, we consider the application model outlined in Chap. 3 in which tasks are activated periodically, scheduled by priority and communicate through shared buffers holding the freshest data values until overwritten by new data (*communication by sampling*).

The approach can be summarized as follows: based on simulation data of a reference CAN bus system, the conjecture is validated that the probability distributions of the message response-times can be approximated by a mix of distribution models with degenerate and gamma distributions. Next, for each message, the correlation between the fitted model, and the message (e.g., size, priority, etc.) and system-level attributes (e.g., bus utilization or the fraction of utilization at higher priority) is computed. The correlation can not only be used to approximate the probability distribution of the response-times for messages belonging to the CAN bus on which the study is performed, but also for other CAN buses. The chapter presents the set of formulae that are found by regression analysis. These can be used to approximate, with good accuracy, the probability distribution of the response-times of messages on CAN buses for which partial and incomplete information is available. Statistical and stochastic analysis (presented in Chap. 4) methods are also compared in terms of speed of the analysis versus availability of data and accuracy of results.

5.2 Deriving and Fitting the Model

Controller Area Network messages are modeled as discussed in Sect. 4.3: a message m_i is defined by the tuple $(T_i, \mathcal{O}_i, \mathcal{E}_i, P_i)$, where T_i is its period, $\mathcal{O}_i$ its initial phase, $\mathcal{E}_i$ its transmission time, and P_i its priority. The jth instance of message m_i as M_{ij}. The pdf of the response-time $f_{\mathcal{R}_i}$ of message m_i is obtained as the average of the response-time pdfs of all the message instances M_{ij}.

We start with the analysis of a reference CAN bus configuration—the messages transmitted over the bus are shown in the Table 5.1. All messages in the system are periodically queued from six nodes. They are indexed in priority order and shown with their periods and transmission times in milliseconds (ms). The speed of the CAN bus is 500 kb/s. For the above set of messages, the utilization of the bus is 60.25%. As for any other CAN-based system, there are three sources of randomness. First, there is lack of clock synchronization among nodes. Clock drifts result in variable offsets for the times at which messages are transmitted. The amount of interference suffered by each message depends on the time at which higher priority messages are contending for the network from remote nodes and is therefore dependent on the offsets. Second, the message load can be variable; however, we target buses in which the message load consists of periodic messages, with loads of constant length. This is representative of a quite significant fraction of bus systems in control and automotive applications (in the experimental vehicles that

Table 5.1 An example automotive CAN system

Msg	ECU	T_i	$\mathcal{E}_i$	P_i	Msg	ECU	T_i	$\mathcal{E}_i$	P_i	Msg	ECU	T_i	$\mathcal{E}_i$	P_i
m_1	E_2	10	0.27	1	m_{24}	E_3	25	0.23	24	m_{47}	E_2	50	0.19	47
m_2	E_2	10	0.27	2	m_{25}	E_3	25	0.25	25	m_{48}	E_4	100	0.27	48
m_3	E_3	5	0.19	3	m_{26}	E_2	20	0.27	26	m_{49}	E_3	100	0.27	49
m_4	E_3	10	0.25	4	m_{27}	E_5	25	0.27	27	m_{50}	E_1	100	0.27	50
m_5	E_1	10	0.19	5	m_{28}	E_2	20	0.27	28	m_{51}	E_3	100	0.13	51
m_6	E_6	10	0.27	6	m_{29}	E_1	25	0.17	29	m_{52}	E_3	100	0.27	52
m_7	E_1	100	0.27	7	m_{30}	E_2	10	0.21	30	m_{53}	E_1	100	0.19	53
m_8	E_1	100	0.15	8	m_{31}	E_2	20	0.27	31	m_{54}	E_3	100	0.13	54
m_9	E_1	100	0.17	9	m_{32}	E_4	10	0.27	32	m_{55}	E_3	100	0.13	55
m_{10}	E_5	25	0.27	10	m_{33}	E_3	10	0.27	33	m_{56}	E_5	100	0.27	56
m_{11}	E_1	100	0.15	11	m_{34}	E_3	10	0.27	34	m_{57}	E_3	100	0.25	57
m_{12}	E_1	20	0.19	12	m_{35}	E_1	25	0.25	35	m_{58}	E_3	100	0.13	58
m_{13}	E_2	100	0.19	13	m_{36}	E_6	25	0.23	36	m_{59}	E_3	100	0.13	59
m_{14}	E_1	100	0.19	14	m_{37}	E_4	50	0.27	37	m_{60}	E_4	100	0.17	60
m_{15}	E_5	100	0.27	15	m_{38}	E_4	50	0.27	38	m_{61}	E_1	100	0.27	61
m_{16}	E_1	10	0.23	16	m_{39}	E_5	50	0.27	39	m_{62}	E_1	100	0.13	62
m_{17}	E_2	100	0.25	17	m_{40}	E_3	50	0.21	40	m_{63}	E_3	100	0.19	63
m_{18}	E_2	100	0.27	18	m_{41}	E_4	50	0.27	41	m_{64}	E_1	100	0.27	64
m_{19}	E_2	50	0.25	19	m_{42}	E_6	50	0.27	42	m_{65}	E_1	100	0.13	65
m_{20}	E_3	10	0.27	20	m_{43}	E_2	50	0.27	43	m_{66}	E_1	100	0.13	66
m_{21}	E_6	10	0.27	21	m_{44}	E_5	100	0.27	44	m_{67}	E_2	50	0.27	67
m_{22}	E_6	25	0.27	22	m_{45}	E_2	25	0.15	45	m_{68}	E_1	100	0.13	68
m_{23}	E_3	25	0.23	23	m_{46}	E_2	50	0.19	46	m_{69}	E_4	100	0.27	69

are the subject of our analysis all but one fall under this category). The third source of variability (of lesser importance) considered in our simulations is bit stuffing because of the variable data content of messages.

5.2.1 Common Characteristics of Message Response-Times

As a first step, a simulator is used to extract the distribution of the response-time values for all messages in our reference system (Table 5.1). The hyperperiod of the system (least common multiple of message periods) is 100 ms. The simulation is performed with a granularity of 0.05 ms, leading to $2{,}000^5 = 3.2 \times 10^{18}$ possible combinations of relative phases among nodes. This space cannot be explored exhaustively and the simulation is performed by randomly choosing relative phases, and repeated for 2×10^{10} times, which takes about 100 h. We used 0.01 ms (or equivalently 10 μs) as the default unit for all time-related quantities. The simulator used for the experiments is a custom-developed (in C++ code) frame-level CAN bus simulator [65] with a generator of randomized data content for each message (of fixed length). The simulator assumes priority-based queuing of messages at each

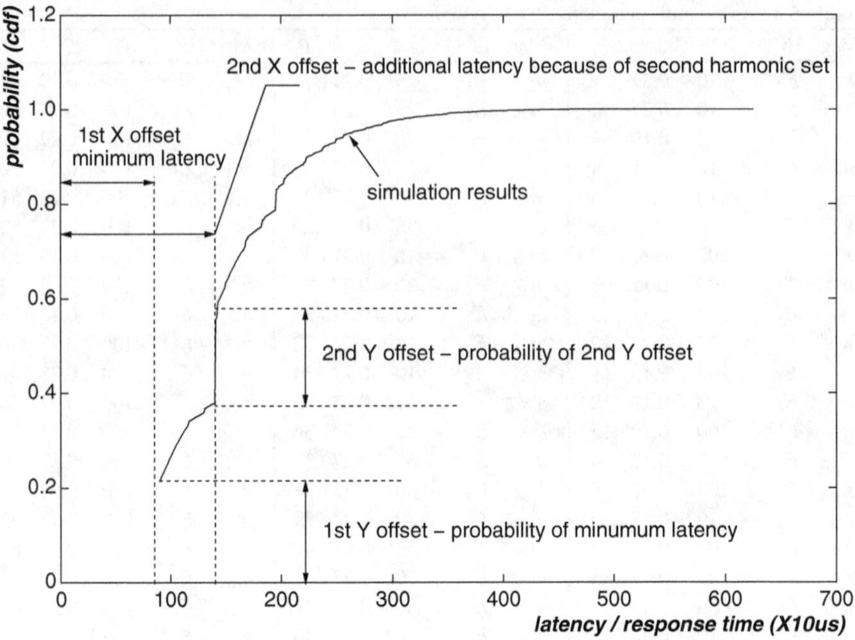

Fig. 5.1 Response-time cdf of m_{25} with more than one higher priority harmonic set

node and availability of a sufficient number of TxObjects at the adapter to avoid priority inversion (see Chap. 3). It also assumes a negligible copy time between the message queue and the adapter registers, and the possibility of aborting message transmissions if a higher priority message becomes ready. The bus contention is simulated according to the CAN 2.0 standard, and simulation of transmission includes bit stuffing.

In our system model, as in most real systems, there is a single middleware task at each node that is responsible for message enqueuing, running at the greatest common divider of the message periods. This means that a low priority message is typically always enqueued together with higher priority messages on the same node. This is indeed the architecture of one of the most used among all the CAN Interaction Layers [4], as well as the pattern that is enforced by code generators (like TargetLink by dSPACE or Embedded Coder by The MathWorks) when the CAN bus is accessed by a custom blockset in a model-based design flow (as in the popular Autobox suite by dSPACE).

The simulation results show that the distributions of message response-times are of different types but with some common characteristics. The type of distribution depends on the message priority, on the characteristics (period, size) of higher priority messages on the local node, and possibly on other (unknown) factors. For example, the *cumulative distribution function* (cdf) of message m_{25} (priority 25 out of 69) is shown in Fig. 5.1. m_{25} is a medium-priority message on the node E_3 which has five messages with higher priority than m_{25}. It is always enqueued together with three higher priority messages, $\{m_3, m_{23}, m_{24}\}$, while other higher priority messages $\{m_4, m_{20}\}$, with period 10 ms, are only enqueued every two instances of m_{25}. We label the set of messages that are always enqueued together with a generic message m_i on a node N_j as the *first local harmonic higher priority set* or *the first harmonic set of level* P_i *for node j*. Similarly, the set of messages that are enqueued together with every second instance of m_i is labeled as the *second local harmonic higher priority set*. In the example of m_{25}, the first local harmonic higher priority set is $\{m_3, m_{23}, m_{24}\}$, and the second harmonic higher priority set is $\{m_3, m_{23}, m_{24}\} \cup \{m_4, m_{20}\}$.

For message m_i, the smallest possible response-time value is the time it takes to transmit all the messages in its first harmonic set of level P_i, plus itself, without any interference or blocking from remote load. This minimum response-time value is labeled as the first *X offset* in Fig. 5.1 and has a finite probability, resulting in a step of the cdf (the first *Y offset* in the figure). Furthermore, the message cdf may have other discontinuities for other higher priority harmonic sets. These discontinuities are characterized by other pairs of X and Y offsets. For example, message m_{25} has three messages in its first harmonic higher priority set. The probability that these three messages and m_{25} are transmitted without interference is approximately 0.2. m_{25} has two other messages in the second harmonic higher priority set, resulting in a second pair of X and Y offsets.

Other messages with different priority may exhibit different characteristics of the cdf. The conjecture is that the message response-time cdf, normalized by shifting left and down along the axes until the first set of X and Y offsets are zero, can be approximated by fitting, using a gamma distribution whose shape parameter a and scale parameter b can be expressed as functions of simple (aggregate) attributes of the message set. In technical terms, the probability mix model of the degenerate distribution and the gamma distribution is used to fit the message response-times.

5.2.2 Fitting the Message Response-Times

The pmf and cdf of a degenerate distribution, which is the probability distribution of a discrete random variable whose support consists of only one value x^{off}, are given by

$$f^D\left(x, x^{\mathrm{off}}\right) = \begin{cases} 0 & \text{if } x \neq x^{\mathrm{off}} \\ 1 & \text{if } x = x^{\mathrm{off}} \end{cases}, \tag{5.1}$$

$$F^D\left(x, x^{\mathrm{off}}\right) = \begin{cases} 0 & \text{if } x < x^{\mathrm{off}} \\ 1 & \text{if } x \geq x^{\mathrm{off}} \end{cases}. \tag{5.2}$$

The pdf of a gamma distribution has the form

$$f^{\Gamma}(x,a,b) = \begin{cases} 0 & \text{if } x \le 0 \\ x^{a-1}\dfrac{e^{-x/b}}{b^{a}\Gamma(a)} & \text{if } x > 0 \end{cases}, \tag{5.3}$$

where b is the scale parameter, a is the shape parameter, and $\Gamma(\cdot)$ is the gamma function $\Gamma(a) = \int_0^{\infty} t^{a-1}e^{-t}\mathrm{d}t$. Its cdf is expressed using the incomplete gamma function $\gamma(a,x) = \int_0^{x} t^{a-1}e^{-t}\mathrm{d}t$,

$$F^{\Gamma}(x,a,b) = \begin{cases} 0 & \text{if } x < 0 \\ \dfrac{\gamma(a,x/b)}{\Gamma(a)} & \text{if } x \ge 0 \end{cases}. \tag{5.4}$$

If we shift the gamma distribution to the right by x^{off}, the cdf of the *offsetted gamma distribution* becomes

$$F^{\Gamma}(x,x^{\text{off}},a,b) = \begin{cases} 0 & \text{if } x < x^{\text{off}} \\ \dfrac{\gamma(a,(x-x^{\text{off}})/b)}{\Gamma(a)} & \text{if } x \ge x^{\text{off}} \end{cases}. \tag{5.5}$$

The response-time of message m_i is assumed to be a random variable whose probability distribution is a probability mix model of a (discrete) degenerate and a (continuous) offset gamma distribution

$$\begin{aligned} & F_i\left(x,x_i^{\text{off}},a_i,b_i,y_i^{D},y_i^{\Gamma}\right) \\ &= y_i^{D}F^{D}\left(x,x_i^{\text{off}}\right) + y_i^{\Gamma}F^{\Gamma}\left(x,x_i^{\text{off}},a_i,b_i\right), \\ &= \begin{cases} 0 & \text{if } x < x_i^{\text{off}} \\ y_i^{D} + y_i^{\Gamma}\dfrac{\gamma(a_i,(x-x_i^{\text{off}})/b_i)}{\Gamma(a_i)} & \text{if } x \ge x_i^{\text{off}} \end{cases}, \\ & \text{w.r.t. } y_i^{D} + y_i^{\Gamma} = 1 \\ & \qquad 0 \le y_i^{D}, y_i^{\Gamma} \le 1 \end{aligned} \tag{5.6}$$

where x_i^{off}, the minimum message response-time, is calculated as the sum of the message transmission time and the queuing delay from local higher priority messages, and the parameters $\left(a_i,b_i,y_i^{D},y_i^{\Gamma}\right)$ are estimated using the expectation-maximization (EM) algorithm. The result of the fit to the response-time of message m_{13} in the set, which has only one local higher priority harmonic set, is shown in Fig. 5.2. The fitted distribution approximation appears to be sufficiently accurate for early estimations.

For messages with more than one local higher priority harmonic set, the response-time cdf is approximated as the sum of multiple pairs of degenerate

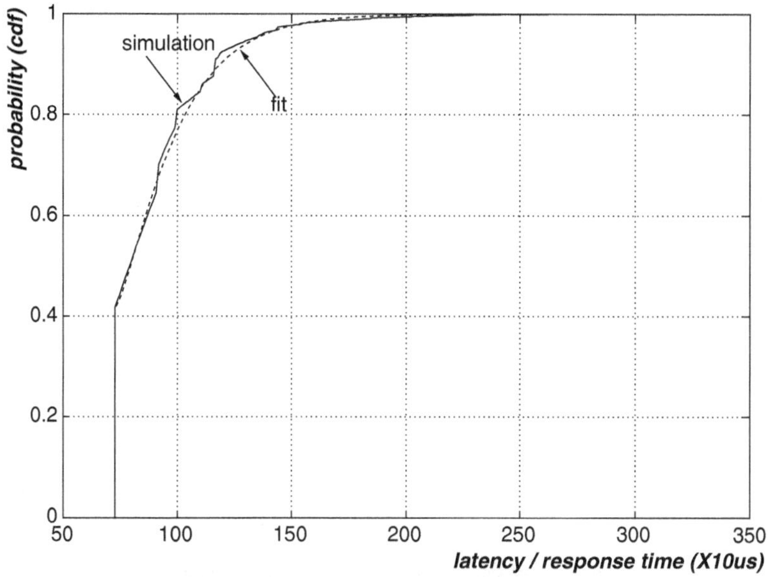

Fig. 5.2 Fitting a mix model distributions to the response-time of m_{13}

and offsetted gamma distributions, each with six characteristic parameters $\left(x_{i,k}^{\text{off}}, y_{i,k}, a_{i,k}, b_{i,k}, y_{i,k}^{D}, y_{i,k}^{\Gamma}\right)$, as below

$$
\begin{aligned}
& F_i\left(x, \bigcup_k \left(x_{i,k}^{\text{off}}, a_{i,k}, b_{i,k}, y_{i,k}^{D}, y_{i,k}^{\Gamma}\right)\right) \\
& = \sum_k F_{i,k}\left(x, x_{i,k}^{\text{off}}, a_{i,k}, b_{i,k}, y_{i,k}^{D}, y_{i,k}^{\Gamma}\right), \\
& \text{where } F_{i,k}\left(x, x_{i,k}^{\text{off}}, a_{i,k}, b_{i,k}, y_{i,k}^{D}, y_{i,k}^{\Gamma}\right) \\
& = y_{i,k}^{D} F^{D}\left(x, x_{i,k}^{\text{off}}\right) + y_{i,k}^{\Gamma} F^{\Gamma}\left(x, x_{i,k}^{\text{off}}, a_{i,k}, b_{i,k}\right), \\
& = \begin{cases} 0 & \text{if } x < x_{i,k}^{\text{off}} \\ y_{i,k}^{D} + y_{i,k}^{\Gamma} \dfrac{\gamma\left(a_{i,k}, \left(x - x_{i,k}^{\text{off}}\right)/b_{i,k}\right)}{\Gamma\left(a_{i,k}\right)} & \text{if } x \geq x_{i,k}^{\text{off}} \end{cases}. \\
& \text{w.r.t. } y_{i,k}^{D} + y_{i,k}^{\Gamma} = y_{i,k}, \forall k \\
& \quad 0 \leq y_{i,k}^{D}, y_{i,k}^{\Gamma} \leq y_{i,k}, \forall k
\end{aligned}
\tag{5.7}
$$

Note that the offset $x_{i,k}^{\text{off}}$ is the total transmission time of the local higher priority harmonic set plus itself, and $y_{i,k}^{D}$ is the probability of its occurrence, considering only the local message set. These two parameters can be estimated by analyzing the local message set. Of course, if the set of messages transmitted by the node is not

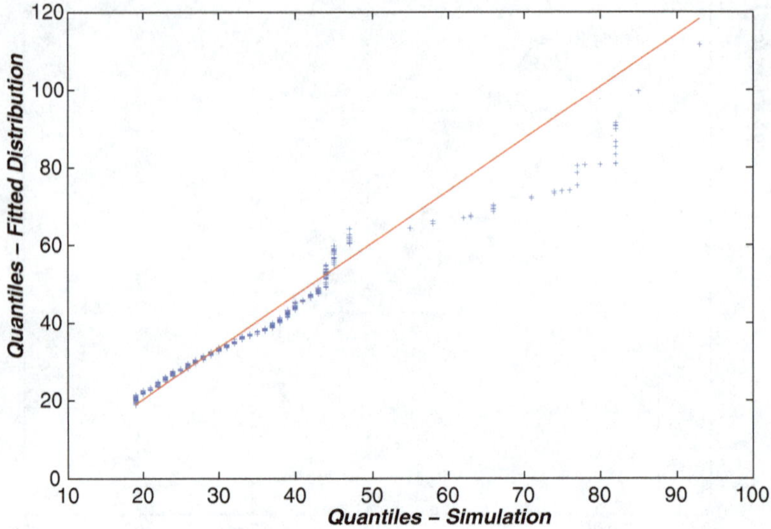

Fig. 5.3 Quantile–quantile plot of samples from simulation and fitted distribution for m_5

fully known, it needs to be approximated with the expected number of messages in the harmonic set, and then make assumptions on the size of these messages (messages tend to be of maximum size because of network efficiency reasons).

The Quantile-Quantile plot (Q–Q plot) is used to graphically compare the probability distributions from the simulation data and the fitted mix model. A Q–Q plot is generated for two distributions D_1 and D_2 by plotting their quantiles against each other. If the two distributions being compared are similar, the points in the Q–Q plot will approximately lie on the line $D_2 = D_1$ (red/solid line in Figs. 5.3 and 5.4). One thousand random samples from each are taken, and the results for message m_5 and m_{69} (high and low priority cases) are shown in Figs. 5.3 and 5.4 respectively. The Q–Q plots for these messages are approximately linear, with higher relative accuracy in the case of low priority messages like m_{69}.

Also, to quantitatively evaluate our hypothesis that the message response-time cdf can be approximated by one or multiple pairs of degenerate and offsetted gamma distributions, several metrics are presented to assess the accuracy of the fitted distribution using the EM algorithm compared to the simulation data: the root mean squared error (RMSE), the coefficient of determination R^2, and the Kolmogorov–Smirnov statistic (K–S statistic, column "K–S", the maximum vertical deviation between the two cdf curves). The results for the messages of the reference bus are shown in Table 5.2. The maximum RMSE is 0.034, the minimum coefficient of determination is 0.980, and the maximum K–S statistic is 0.11, all indicating that our hypothesis is accurate for the purpose of an early estimation.

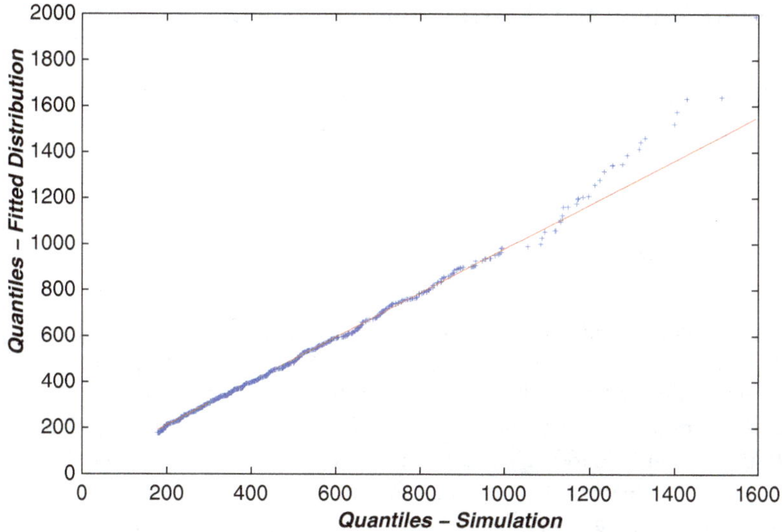

Fig. 5.4 Quantile–quantile plot of samples from simulation and fitted distribution for m_{69}

Table 5.2 Statistics of the fitted model distributions for messages on the reference bus

ID	RMSE	R^2	K–S	ID	RMSE	R^2	K–S	ID	RMSE	R^2	K–S
1	0.017	0.985	0.03	24	0.015	0.994	0.07	47	0.031	0.988	0.07
2	0.017	0.985	0.03	25	0.016	0.994	0.08	48	0.008	0.998	0.02
3	0.009	0.994	0.03	26	0.012	0.994	0.04	49	0.007	0.999	0.03
4	0.012	0.989	0.03	27	0.003	1.000	0.02	50	0.014	0.997	0.04
5	0.009	0.995	0.04	28	0.013	0.995	0.04	51	0.008	0.999	0.03
6	0.008	0.995	0.03	29	0.020	0.989	0.06	52	0.010	0.998	0.03
7	0.009	0.995	0.04	30	0.011	0.998	0.04	53	0.015	0.997	0.04
8	0.010	0.996	0.04	31	0.014	0.996	0.04	54	0.011	0.998	0.04
9	0.010	0.996	0.04	32	0.008	0.993	0.04	55	0.013	0.997	0.04
10	0.009	0.994	0.04	33	0.006	0.999	0.03	56	0.008	0.998	0.02
11	0.010	0.997	0.04	34	0.007	0.999	0.03	57	0.014	0.997	0.05
12	0.021	0.983	0.06	35	0.018	0.993	0.05	58	0.016	0.997	0.05
13	0.008	0.996	0.04	36	0.008	0.998	0.04	59	0.017	0.996	0.05
14	0.011	0.997	0.05	37	0.005	0.999	0.02	60	0.006	0.999	0.02
15	0.010	0.994	0.05	38	0.007	0.998	0.02	61	0.012	0.998	0.03
16	0.015	0.993	0.05	39	0.006	0.999	0.01	62	0.013	0.998	0.03
17	0.008	0.997	0.05	40	0.006	0.999	0.02	63	0.015	0.997	0.05
18	0.009	0.997	0.04	41	0.009	0.998	0.02	64	0.012	0.998	0.03
19	0.034	0.981	0.11	42	0.009	0.998	0.03	65	0.013	0.998	0.03
20	0.007	0.994	0.05	43	0.030	0.986	0.09	66	0.012	0.998	0.03
21	0.010	0.991	0.06	44	0.010	0.998	0.03	67	0.018	0.995	0.05
22	0.017	0.992	0.08	45	0.029	0.980	0.07	68	0.012	0.999	0.03
23	0.014	0.994	0.07	46	0.030	0.988	0.08	69	0.003	1.000	0.02

5.3 Estimate of the Distribution Parameters

In order to reconstruct the response-time probability distribution of a message m_i for the model in (5.7), we need to estimate the parameters $\left(x^{\text{off}}_{i,k}, y_{i,k}, a_{i,k}, b_{i,k}, y^D_{i,k}, y^{\Gamma}_{i,k}\right)$.

5.3.1 *Parameters x^{off} and y*

$x^{\text{off}}_{i,k}$ and $y_{i,k}$ can be calculated by considering the interference from the local higher priority message set. Consider, for example, message m_{39} with period 5,000 and transmission time 27 on node E_5. There are three higher priority messages on the same node, m_{10}, m_{15}, and m_{27}, with period and transmission times (2,500, 27), (10,000, 27), and (2,500, 27), respectively. During the hyperperiod [0,10 000), message m_{39} is queued at time 0 and 5,000. At time 0, all three higher priority messages m_{10}, m_{15}, and m_{27} are queued, thus the minimum response-time of the message instance from m_{39} is 108. At time 5,000, m_{10} and m_{27} are queued together with m_{39}, the minimum response-time of m_{39} is 81. Due to the different minimum response-times of these two instances from m_{39}, there are two pairs of $x^{\text{off}}_{i,k}$ and $y_{i,k}$ parameters: ($x^{\text{off}}_{39,1} = 81, y_{39,1} = 0.5$) and ($x^{\text{off}}_{39,2} = 108, y_{39,2} = 0.5$).

5.3.2 *Parameters y^D and y^{Γ}*

Since in our model (5.7), the constraint $y^D_{i,k} + y^{\Gamma}_{i,k} = y_{i,k}$ must be satisfied, we obtain $y^{\Gamma}_{i,k} = y_{i,k} - y^D_{i,k}$ after estimating $y_{i,k}$ and $y^D_{i,k}$. In the following, we focus on $y^D_{i,k}$, the probability that all the messages in the corresponding higher priority harmonic set and m_i itself are transmitted immediately upon enqueuing, considering the possible interference and blocking from remote load.

We attempted to approximate this stochastic process with a binomial (Bernoulli) process with replacement.[1] The result is a formula that computes $y^D_{i,j}$ as a function of U^r_i, the network utilization from remote nodes, U^{hr}_i, the utilization due to higher priority messages transmitted by other nodes, $y_{i,j}$, the probability of a queuing event with the jth offset (evaluated by considering all the queuing events in the hyperperiod), and $n_{i,j}$, the number of messages in the first higher priority harmonic set for m_i plus one (m_i itself).

$$y^D_{i,j} = \mathbb{P}\left(\mathcal{R}_i = x^{\text{off}}_{i,j}\right) = y_{i,j}\left(1 - U^r_i\right)\left(1 - U^{hr}_i\right)^{n_{i,j}-1}. \tag{5.8}$$

[1] The reader interested in the reasoning behind this model can find details in Ref. [63, pp. 105–108].

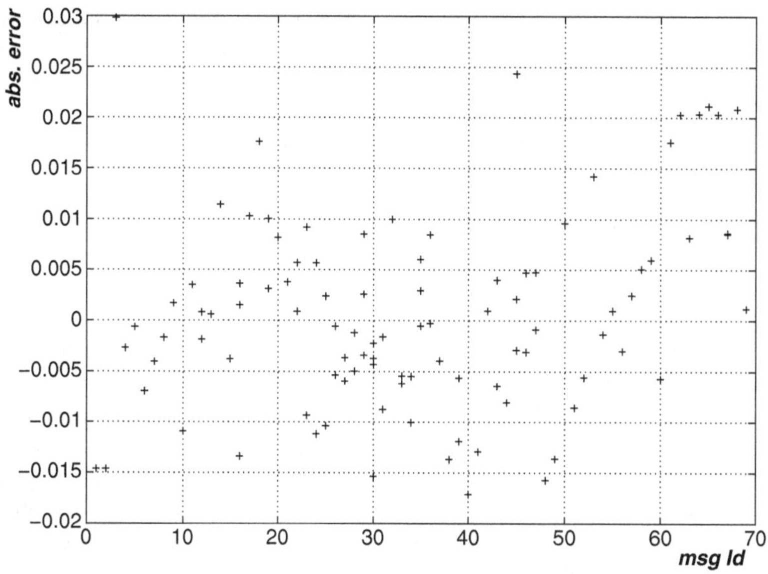

Fig. 5.5 Absolute errors in the estimation of the y^D values for messages

Equation (5.8) is a fair, but not enough accurate approximation. In fact, it is based on a number of simplifying assumptions. The first is that the assignment of the idle/busy status to the bus for the transmission attempts of the higher priority harmonic messages is defined with replacement, while in reality the number of idle/busy slots in the hyperperiod is fixed. More importantly, the assumption that the status of the bus for a transmission attempt is independent from the status at previous attempts leads to an underestimate of the actual probability of long idle times. This is because the synchronization of the transmission times of local messages operated by the middleware task results in bursts of transmissions followed by idle intervals. Hence, the above method provides a good estimate for short queue lengths and an underestimation of the actual probability for longer queues.

However, based on the previous formula, we tried a parameterized model and then used regression to estimate the model parameters. After several attempts, the best model we found in terms of accuracy is the following:

$$
\begin{aligned}
y_{i,j}^D &= \mathbb{P}\left(\mathcal{R}_i = x_{i,j}^{\text{off}}\right) \\
&= y_{i,j}\left(1 - U_i^r\right) e^{-\beta_1 (U_i^{hr} + \beta_2)(Q_{i,j} \times U_i^{hr}/Q_i^{hr} + \beta_3) + \beta_4},
\end{aligned}
\tag{5.9}
$$

where Q_i^{hr} is the quadratic mean of the queueing lengths (expressed as in time) of higher priority messages from remote nodes. Q_i^{hr}/U_i^{hr} represents the average period of the higher priority messages for remote nodes. We define Q_i^{hr}/U_i^{hr} as infinity if there is no remote message with priority higher than that of m_i. Figure 5.5

shows the absolute errors in the evaluation of the probability y^D for all the messages in the set. The regression function (5.9) results in a sufficiently accurate estimate, always within ± 0.03 of the actual probability value, and the RMSE of the estimate is around 0.01.

In [64] we proposed a different regression model:

$$\begin{aligned} y_{i,j}^D &= \mathbb{P}\left(\mathcal{R}_i = x_{i,j}^{\text{off}}\right) \\ &= y_{i,j}\left(1 - U_i^r\right) e^{-\beta_1 (U_i^{hr} + \beta_2)(Q_{i,j}/T_i^{hr} + \beta_3) + \beta_4} \end{aligned} \tag{5.10}$$

which requires knowledge of T_i^{hr}, the minimum greatest common divisor (gcd) of the higher priority messages for remote nodes ($T_i^{hr} = \infty$ if there is no remote message with priority higher than m_i). Intuitively, T_i^{hr} is the minimum distance of two consecutive higher priority message bursts from remote nodes, which is difficult to estimate, especially in presence of uncertainty. A different regression model can be used by replacing T_i^{hr} with Q_i^{hr}/U_i^{hr} and using Q_i^{hr} in the estimates of μ and b (see Sect. 5.3.3). This new regression model is slightly more accurate (equal RMSE and smaller maximum error), but especially, it does not require the additional parameter T_i^{hr}.

The β values in (5.9) are not constant, because y^D depends not only on the predictor parameters $U_i^r, U_i^{hr}, Q_{i,j}$, and Q_i^{hr}, but also on other attributes of the message set. We evaluated other attributes, including the number of nodes in the system, the number of messages for each node, and the average number of harmonic sets for each node. However, the average size of messages is the only one with a significant impact. In CAN, each standard message must be between 0 and 8 bytes. Table 5.3 shows the variation of coefficients $\beta_1 - \beta_4$ in (5.9) for different average message sizes, where $Q_{i,j}$ and Q_i^{hr} are expressed in unit of 10 μs.

The regression formula (9), as well as the following (11)–(14) have been obtained partly by leveraging insight on the characteristics of the system to be modeled (for example, modeling the latency distribution using a Bernoulli process), but mostly they are the result of empirical tuning based on the experimental results. The predictor parameters influencing the distribution are partly obvious (such as U_i^{hr}) and partly have been determined as the result of extensive experiments. Strictly speaking, the determination of the regression formulas is not the result of a systematic method. As in many fitting problems, results are found by trying polynomials of increasing degree (typically not larger than second or third degree) and, when this is not enough, exponentials and eventually functions in which the exponent is a polynomial in the predictor parameters.

5.3.3 Parameters a and b of the Gamma Distribution

The parameters a and b of the gamma distribution in the fitted model are computed for all the messages on the reference bus. The definition of a parameterized model

Table 5.3 Coefficients $\beta_1 - \beta_4$ of the parameterized model of y^D for the reference bus

Size	7.623	6.971	6.478	5.797	5.101	4.377	3.623	3.000	2.406	1.000
β_1	0.9487	0.8240	0.8248	0.8331	0.8247	0.8279	0.9228	1.073	1.013	1.594
β_2	1.282	1.527	1.513	1.475	1.515	1.509	1.285	1.087	1.157	0.6562
β_3	−0.1182	−0.1248	−0.1108	−0.0892	−0.0972	−0.1104	−0.0913	−0.1032	−0.0978	−0.0673
β_4	−0.1676	−0.1702	−0.1434	−0.1079	−0.1155	−0.1294	−0.0973	−0.1105	−0.1027	−0.0637

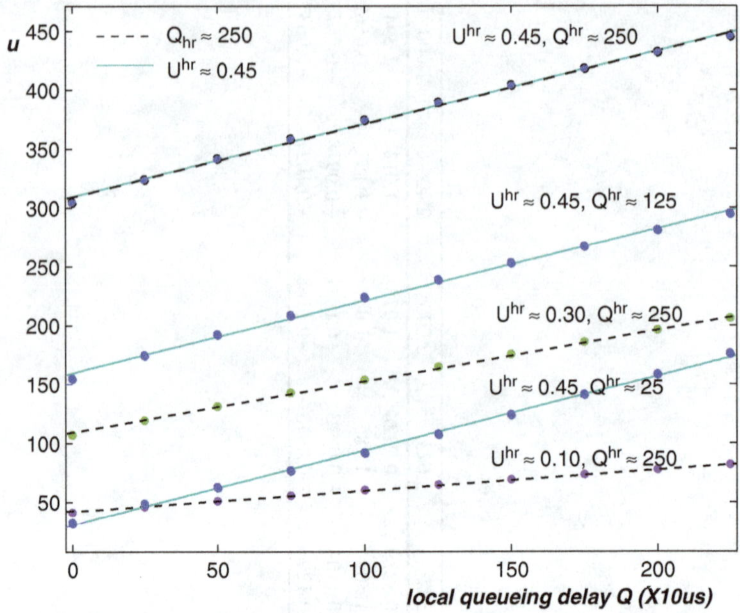

Fig. 5.6 Approximate linear relation between μ and Q

for a and b is made difficult by the absence of a clear physical meaning for them. $\mu_{i,k} = a_{i,k} \times b_{i,k}$, the mean of a gamma distribution with parameters $a_{i,k}$ and $b_{i,k}$, is the average interference delay from remote nodes to m_i when its local queueing delay is $Q_{i,k}$ (in correspondence to $x^{\text{off}}_{i,k}$.) Intuitively, $\mu_{i,k}$ must depend on the local queueing delay $Q_{i,k}$, since the longer $Q_{i,k}$ is, the higher is the probability that m_i will be delayed because of interference from remote messages. $\mu_{i,k}$ should also depend on the higher priority utilization U_i^{hr} from remote notes, which is the fraction of time the bus is busy because of remote higher priority messages. Finally, $\mu_{i,k}$ depends on the quadratic mean Q_i^{hr} of the queueing lengths (expressed as time) of higher priority messages from remote nodes. Q_i^{hr} represents the average time m_i is delayed each time it experiences interference by a burst of messages from one of the remote nodes. In the following, given the factors that affect the value of $\mu_{i,k}$, we define a parameterized model and estimate the fitting coefficients. It turns out the same model can be used for $b_{i,k}$.

5.3.3.1 μ and *b* vs Local Queueing Delay *Q*

To identify the relation between μ and Q, we should find groups of messages that approximately have the same U^{hr} and Q^{hr}, but different Q values. The data extracted from the reference bus were not sufficient, so we created a number of additional configurations by varying the definition of the message set. Figure 5.6

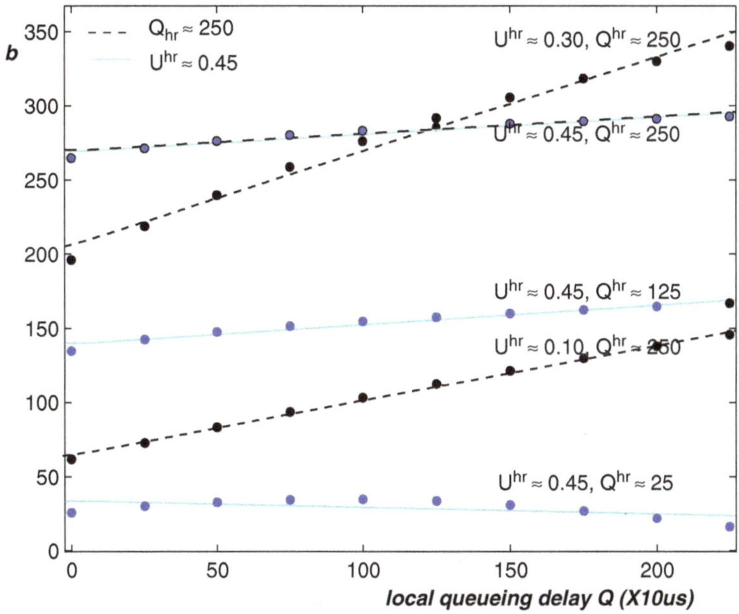

Fig. 5.7 Approximate linear relation between b and Q

plots five groups of data with two sets of pairs (U^{hr}, Q^{hr}). In the first set, the value of U^{hr} is approximately 0.45, and Q^{hr} has values $(25, 125, 250)$. In the second set, $Q^{hr} = 250$ and U^{hr} takes values in $(0.10, 0.30, 0.45)$, respectively. It is clear from the graph that μ is linearly dependent on Q, as all points are closely distributed along the respective linear regression lines. Similarly, Fig. 5.7 illustrates the existence of an approximate linear relation between b and Q for the same sets of messages.

5.3.3.2 μ and b vs Quadratic Mean of Remote Higher Priority Queueing Length Q^{hr}

To find the relation between μ, b and the quadratic mean of the remote higher priority queueing length Q^{hr}, we used four groups of data with (U^{hr}, Q) values around $(0.05, 50)$, $(0.20, 50)$, $(0.45, 50)$, and $(0.45, 225)$. The results of the analysis are shown in Fig. 5.8 for the dependency of μ from Q^{hr}. Figure 5.8 shows an approximate linear dependency between μ and Q^{hr}. A similar dependency exists between b and Q^{hr}, as in Fig. 5.9.

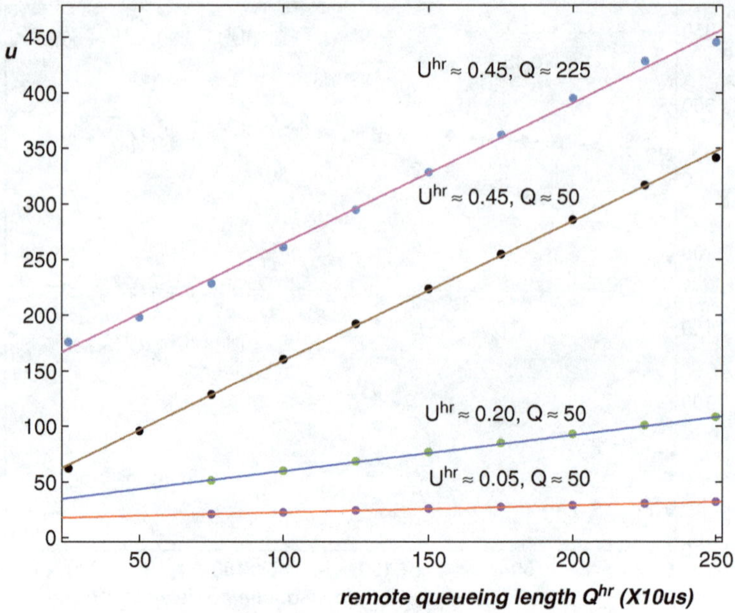

Fig. 5.8 Approximate linear relation between μ and Q^{hr}

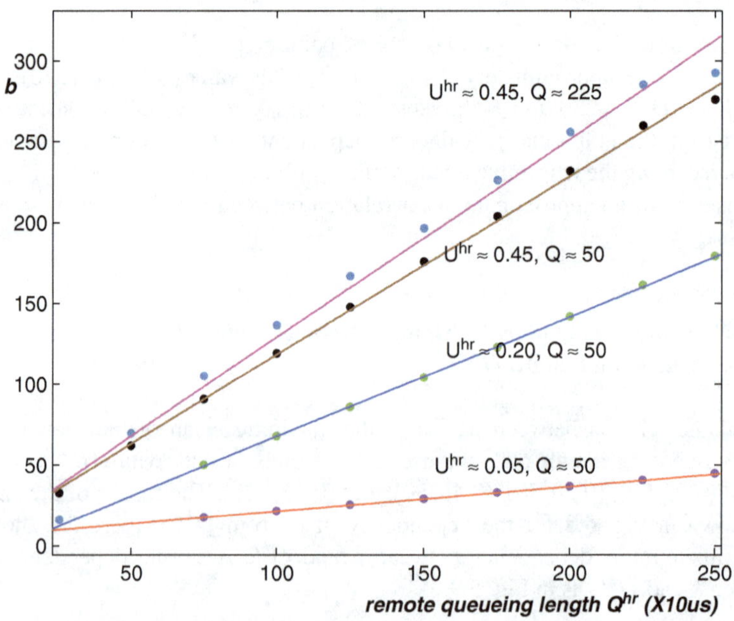

Fig. 5.9 Approximate linear relation between b and Q^{hr}

5.3.3.3 Parameterized Model with Q, Q^{hr}, and U^{hr}

The relation between μ and the predictor parameter U^{hr} is more complex. We leverage the following information

- (from experimental evidence) μ is approximately a linear function of Q. It is also approximately linear with respect to Q^{hr};
- μ, the mean of the gamma distribution, represents a physical time quantity.

Since Q and Q^{hr} are both physical quantities of time, by the theory of quantity calculus, the function expressing μ cannot contain any product (or higher degree combination) of Q and Q^{hr}. Thus, we tried the following parameterized model for μ

$$\mu_{i,k} = Q_{i,k} \times g_1\left(U^{hr}\right) + Q_i^{hr} \times g_2\left(U^{hr}\right) + g_3\left(U^{hr}\right), \tag{5.11}$$

where g_1, g_2, and g_3 are functions of U^{hr}.

Using similar reasoning, it is possible to derive the following model for b:

$$b_{i,k} = Q_{i,k} \times g_4\left(U^{hr}\right) + Q_i^{hr} \times g_5\left(U^{hr}\right) + g_6\left(U^{hr}\right), \tag{5.12}$$

where g_4, g_5, and g_6 are functions of U^{hr}.

At this point, we need to define the functions $g_1 - g_6$ with very little additional knowledge or insight from their physical meaning. In the experiments, we limited our empirical search to polynomial (second degree), exponential and gaussian functions, as well as their sums or products.

Models to Minimize Absolute Error

The parameterized model to minimize the absolute error of μ in the reference bus (as well as other buses, as shown in the following sections) is

$$\mu_{i,k} = \left(Q_{i,k} + \beta_5\right) e^{\beta_6 + \beta_7 U_i^{hr}} + \left(Q_i^{hr} + \beta_8\right) U_i^{hr} e^{\beta_9 + \beta_{10} U_i^{hr}}, \tag{5.13}$$

where $Q_{i,k}$, Q_i^{hr}, and U_i^{hr} are predictor parameters, and $\beta_5 - \beta_{10}$ are coefficients to be determined by regression. The parameterized model for b is a higher order function of U^{hr} with more coefficients:

$$\begin{aligned} b_{i,k} = {} & \left(Q_{i,k} + \beta_{11}\right) e^{\beta_{12} + \beta_{13} U_i^{hr} + \beta_{14} \left(U_i^{hr}\right)^2} \\ & + \left(Q_i^{hr} + \beta_{15}\right) U_i^{hr} e^{\beta_{16} + \beta_{17} U_i^{hr} + \beta_{18} \left(U_i^{hr}\right)^2} + \beta_{19}, \end{aligned} \tag{5.14}$$

where $\beta_{11} - \beta_{19}$ are constant coefficients.

Figure 5.10 shows the absolute errors in the evaluation of the μ and b values for all the messages in the reference set. The regression functions result in a sufficiently accurate estimate, always within ±40 (i.e., ±0.4 ms) of the actual values, and the RMSE of the estimate of μ and b are 12.4 and 6.3 respectively.

Similar to the case of y^D, coefficients $\beta_5 - \beta_{19}$ depend on the average message size. Table 5.4 shows the variation of $\beta_5 - \beta_{19}$ for different message sizes, when $\mu_{i,k}$, $b_{i,k}$, $Q_{i,k}$, and Q_i^{hr} are expressed in 10 μs units.

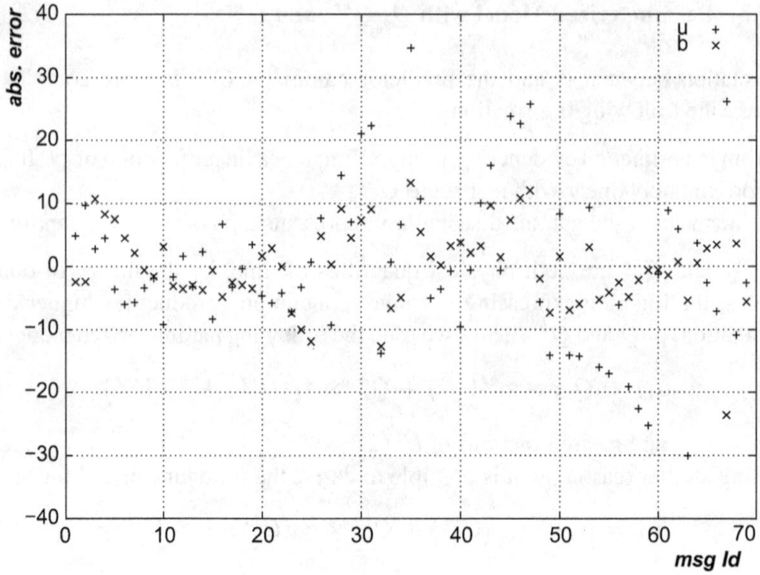

Fig. 5.10 Minimized absolute errors in the estimation of μ and b for messages on the reference bus

Regression functions (5.13) and (5.14) contain 6 and 9 fitting coefficients respectively, and provide results with a good balance of accuracy and complexity. These models have been selected based on the fact that the regression error is significantly higher with a lower number of coefficients, while a function with more coefficients would not bring a significant improvement on the accuracy.

To provide an example, the best model we found for μ using only 4 coefficients is

$$\mu_{i,k} = \left(Q_{i,k} + \alpha_1 Q_i^{hr} + \alpha_2\right) e^{\alpha_3 + \alpha_4 U_i^{hr}}, \tag{5.15}$$

where $\alpha_1 - \alpha_4$ are constant coefficients. This regression function is simpler than (5.13), but less accurate. The average error on μ for all the buses in our experiments increases to 29.6, compared to 22.6 from (5.13).

However, if a higher number of coefficients is used, the best model we found for μ with 8 coefficients is

$$\begin{aligned}\mu_{i,k} &= \left(Q_{i,k} + \alpha_1\right) e^{\alpha_2 + \alpha_3 U_i^{hr}} \\ &\quad + \left(Q_i^{hr} + \alpha_4\right) e^{\alpha_5 + \alpha_6 U_i^{hr}} + \alpha_7 e^{\alpha_8 U_i^{hr}}\end{aligned} \tag{5.16}$$

which can only slightly reduce the error to 22.2.

Similarly, it is possible to identify simpler regression functions for estimating the value of b. For example, the best model we found for b with 6 coefficients is

$$b_{i,k} = \left(Q_{i,k} + \alpha_1\right) e^{\alpha_2 + \alpha_3 U_i^{hr}} + \left(Q_i^{hr} + \alpha_4\right) U_i^{hr} e^{\alpha_5 + \alpha_6 U_i^{hr}} \tag{5.17}$$

Table 5.4 Coefficients $\beta_5 - \beta_{19}$ of the parameterized model to minimize absolute errors of μ and b

Size	7.623	6.971	6.478	5.797	5.101	4.377	3.623	3.000	2.406	1.000
β_5	93.74	101.3	96.49	95.11	64.63	66.34	50.67	64.71	66.05	45.76
β_6	−1.470	−1.635	−1.700	−1.719	−1.655	−1.719	−1.682	−1.901	−1.999	−2.031
β_7	2.453	2.615	2.659	2.561	2.249	2.119	1.780	2.010	2.188	1.561
β_8	−86.30	−78.80	−71.98	−66.77	−50.03	−44.17	−33.50	−31.37	−28.63	−17.21
β_9	1.025	1.023	0.9557	1.001	0.7454	1.115	1.107	1.631	1.669	1.940
β_{10}	1.472	1.191	1.133	1.042	1.465	0.8626	0.8542	−0.3091	−0.4888	−1.272
β_{11}	−22.27	−24.88	6.405	−16.96	−28.13	−46.19	−45.22	8.214	47.86	108.1
β_{12}	−11.00	−7.414	−5.688	−6.809	−6.568	−5.235	−5.111	−4.650	−3.056	−3.036
β_{13}	47.89	29.96	21.92	28.33	29.45	23.65	23.63	21.70	7.622	4.237
β_{14}	−60.03	−38.78	−30.12	−38.97	−44.88	−40.84	−44.32	−47.34	−19.65	−8.809
β_{15}	−31.40	−25.58	−25.05	−20.68	−20.36	−17.84	−18.29	−21.19	−21.97	−17.39
β_{16}	3.213	2.841	2.614	2.569	2.628	2.527	2.663	2.770	2.631	3.232
β_{17}	−8.936	−7.602	−7.160	−7.031	−7.532	−6.876	−7.477	−7.888	−6.574	−13.14
β_{18}	10.73	9.572	9.590	9.384	10.51	10.03	11.25	12.72	10.73	27.11
β_{19}	6.215	4.223	4.254	3.274	4.067	3.336	3.344	3.419	0.1414	−3.092

However, this new regression function will result in a larger average error for b by 55.6% compared to that of (5.14).

Models to Minimize Relative Error

Equations (5.13) and (5.14) provide estimates with small absolute errors. Larger relative errors are obtained for high priority, low latency messages, as their μ and b values are typically small (≤ 40), and the predicted distribution deviates significantly from the one obtained by simulation. If the objective is to minimize relative errors in the estimation on μ and b, a better parameterized model of μ is

$$\mu_{i,k} = Q_{i,k} U_i^{hr} e^{\beta'_5 + \beta'_6 U_i^{hr}} + Q_i^{hr} U_i^{hr} e^{\beta'_7 + \beta'_8 U_i^{hr}} \\ + \beta'_9 e^{\beta'_{10} U_i^{hr}} + \beta'_{11} U_i^{hr} e^{\beta'_{12} U_i^{hr}}. \tag{5.18}$$

Correspondingly, the model for b has the same format as in (5.14) but with a different set of coefficients β'. The values to be assigned to $\beta'_5 - \beta'_{21}$ depend on the average message size. Table 5.5 shows the estimates of $\beta'_5 - \beta'_{21}$ for different message sizes.

$$b_{i,k} = \left(Q_{i,k} + \beta'_{13}\right) e^{\beta'_{14} + \beta'_{15} U_i^{hr} + \beta'_{16} \left(U_i^{hr}\right)^2} \\ + \left(Q_i^{hr} + \beta'_{17}\right) U_i^{hr} e^{\beta'_{18} + \beta'_{19} U_i^{hr} + \beta'_{20} \left(U_i^{hr}\right)^2} + \beta'_{21}. \tag{5.19}$$

Figure 5.11 shows the relative errors in the evaluation of the μ and b values for all the messages in the reference set. The relative error from the regression functions (5.18) and (5.19) are always within $\pm 25\%$, and the RMSE of the estimate of μ and b are 8.1% and 8.3% respectively.

Table 5.4 summarizes the variation of $\beta'_5 - \beta'_{21}$ for different message sizes, where $\mu_{i,k}$, $b_{i,k}$, $Q_{i,k}$, and Q_i^{hr} are expressed in 10 μs units.

5.4 Predicting Message Response-Times

One of the goals of the statistical analysis is to verify if the methods for estimating the X offsets, the y^D and y values, and the a and b parameters of the gamma distributions assumed as an approximation of the message response-time cdf can be used to predict the response-time distribution of a given message, either on the reference bus, or on other buses. The results are shown in the following figures, where the curve "simulation" is the message response-time cdf from simulation data, "fit" is the curve using the X offsets and y values calculated from Sect. 5.3.1 combined with the y^D, a, and b parameters from fitting the model distribution to simulation data, "prediction" is the one obtained by using the X offsets and y values from Sect. 5.3.1 combined with the y^D parameter and μ, a, b parameters

Table 5.5 Coefficients $\beta'_5 - \beta'_{21}$ of the parameterized model to minimize relative errors of μ and b

Size	7.623	6.971	6.47	5.797	5.101	4.377	3.623	3.000	2.406	1.000
β'_5	−0.2162	−0.1471	−0.0894	−0.0815	0.0819	0.2145	0.3537	0.3988	0.4723	0.4972
β'_6	1.048	0.7659	0.5758	0.4898	$-1.2E-4$	−0.5215	−1.088	−1.562	−1.983	−2.583
β'_7	0.1656	0.2159	0.2634	0.2885	0.6368	0.7923	1.027	1.383	1.317	2.061
β'_8	2.951	2.986	2.812	2.706	1.730	1.644	0.9226	0.4170	0.9679	−2.498
β'_9	13.02	12.29	12.04	11.86	11.26	10.21	9.550	8.678	7.703	6.399
β'_{10}	7.486	7.680	7.707	7.700	7.642	7.783	7.696	7.747	8.953	8.683
β'_{11}	−90.75	−87.50	−86.93	−86.11	−94.45	−97.14	−104.0	−118.3	−105.3	−142.9
β'_{12}	5.420	5.569	5.526	5.485	5.020	4.908	4.358	3.913	5.075	2.163
β'_{13}	194.0	229.7	205.1	175.4	175.5	138.1	172.1	200.7	152.6	252.1
β'_{14}	−3.466	−3.310	−3.418	−3.247	−3.375	−3.363	−3.281	−3.630	−3.464	−3.525
β'_{15}	9.611	8.639	9.799	8.857	9.204	10.50	9.159	11.97	13.18	9.921
β'_{16}	−13.55	−12.89	−15.49	−14.26	−14.95	−19.94	−18.28	−27.98	−35.81	−28.03
β'_{17}	−49.80	−50.05	−47.91	−44.13	−39.29	−36.45	−35.00	−32.75	−31.13	−23.78
β'_{18}	1.551	1.214	1.142	1.029	1.745	1.718	2.017	2.146	1.749	2.772
β'_{19}	−1.804	0.2054	0.6281	1.199	−3.197	−2.488	−3.982	−3.336	−0.2080	−10.12
β'_{20}	3.817	1.690	1.250	0.3124	6.140	5.535	7.904	6.616	1.834	24.68
β'_{21}	−0.5395	−3.300	−1.712	−1.866	−1.120	−0.5272	−2.478	−1.440	−1.796	−4.905

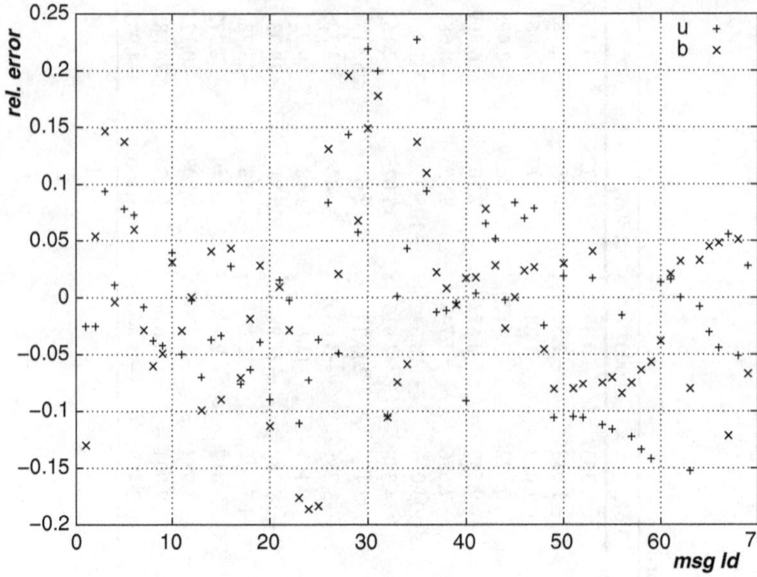

Fig. 5.11 Minimized relative errors in the estimation of μ and b for messages on the reference bus

from the regression functions. In the following, we use (5.9), (5.13), and (5.14), i.e., to minimize the absolute errors of μ and b. The latter two equations can also be replaced by (5.18) and (5.19) to minimize the relative errors of μ and b.

5.4.1 Prediction of Response-Time cdfs for Messages on the Reference Bus

Of course, the accuracy of the predictions of the statistical analysis and the quality of the proposed regression formulae must be verified by the application to a number of case studies. In the first experiment, we verify the capability of predicting the response-time of a message on the reference bus.

The first plot, in Fig. 5.12, shows the results for a high priority message. The predicted parameters y^D and μ, a, b are close to the ones of the fitted model, which results in a quite accurate approximation of the actual cdf. The second case, in Fig. 5.13, is a representative of low priority messages. For this message, the prediction is still very close to the simulated data. In general, for all the messages in the set, the error is acceptable for an early design estimation. As shown in the figures, the quality of the approximation does not necessarily depend on the priority of the message but rather on the shape of the cdf.

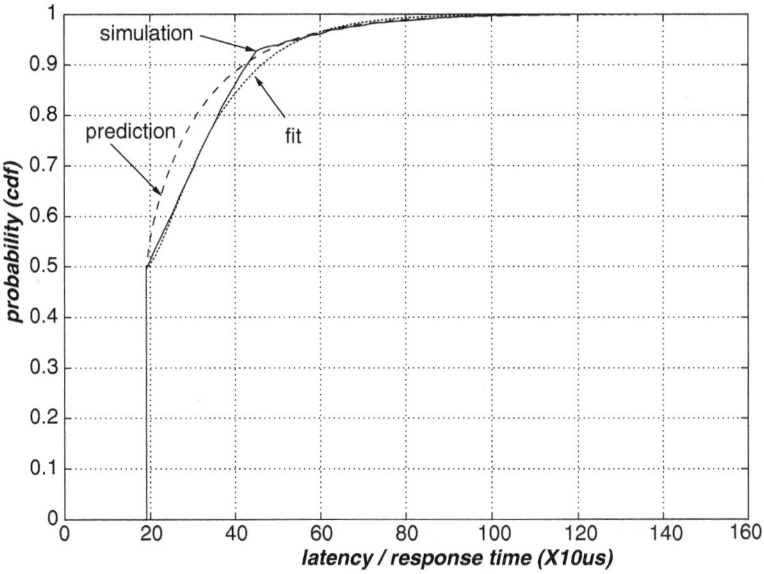

Fig. 5.12 Prediction of the response-time cdf for message m_5

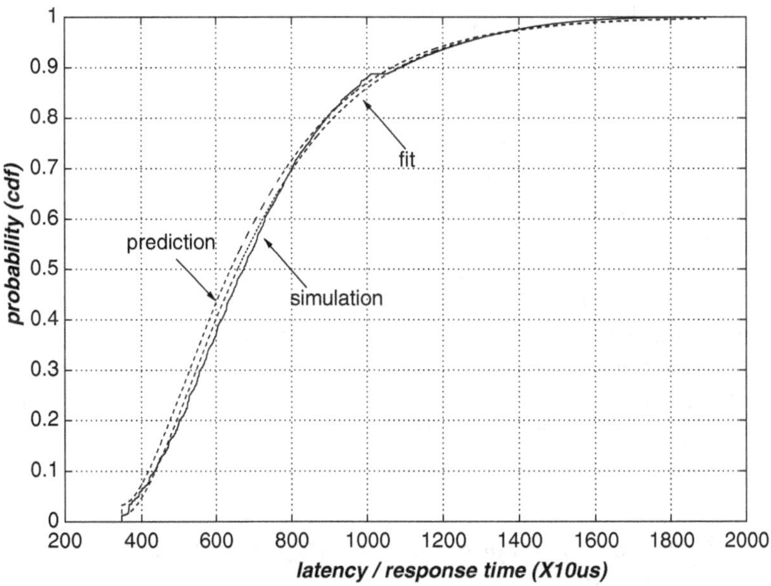

Fig. 5.13 Prediction of the response-time cdf for message m_{68}

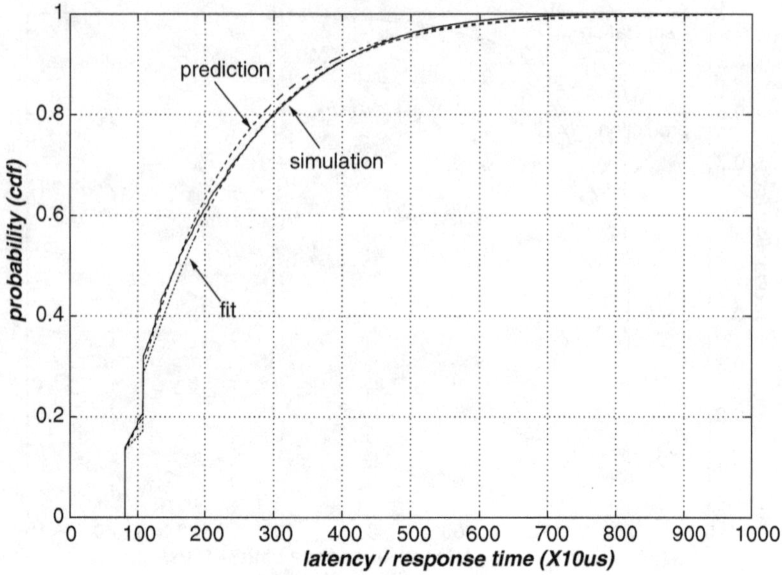

Fig. 5.14 Prediction of response-time cdf for m_{39} with more than one harmonic set

The method also allows to approximate messages with more than one higher priority harmonic set, as shown in the sample case of Fig. 5.14, which refers to the medium-priority message m_{39}.

The root mean squared errors, the coefficients of determination R^2, and the K–S statistics for messages on the reference bus are shown in Table 5.6. Other statistics are also accurate enough for an early stage estimation. For example, the relative error in the evaluation of the average response-time for all the messages is always below 12.4%.

Table 5.6 is based on the regression functions in (5.13) and (5.14), which are to minimize the absolute error of μ and b. As expected, the relative quality of estimation for high priority messages is in general worse than those for low priority messages. Equations (5.18) and (5.19) can be used to obtain better results with respect to relative errors (as opposed to absolute errors). The results when using (5.18) and (5.19) are summarized in Table 5.7.

5.4.2 *Prediction of Response-Time cdfs for Messages on Other Buses*

The $\beta_1 - \beta_{19}$ regression coefficients of (5.9), (5.13), and (5.14) are estimated on the reference bus and then used to predict the response-time cdf of messages on other buses. Messages on other buses of the same vehicle as well as other vehicles are used

Table 5.6 Error of the predicted distribution for messages on the reference bus with regression functions (5.9), (5.13), and (5.14)

ID	RMSE	R^2	K–S	ID	RMSE	R^2	K–S	ID	RMSE	R^2	K–S
1	0.033	0.932	0.07	24	0.013	0.996	0.05	47	0.021	0.994	0.05
2	0.058	0.852	0.11	25	0.012	0.997	0.05	48	0.010	0.998	0.04
3	0.028	0.944	0.06	26	0.017	0.989	0.06	49	0.016	0.995	0.04
4	0.031	0.925	0.07	27	0.015	0.990	0.04	50	0.012	0.998	0.03
5	0.032	0.935	0.10	28	0.017	0.992	0.06	51	0.017	0.994	0.05
6	0.038	0.896	0.13	29	0.009	0.998	0.03	52	0.019	0.994	0.06
7	0.031	0.948	0.09	30	0.019	0.992	0.06	53	0.013	0.997	0.03
8	0.018	0.987	0.05	31	0.018	0.993	0.07	54	0.022	0.992	0.07
9	0.013	0.994	0.05	32	0.017	0.972	0.04	55	0.025	0.990	0.08
10	0.046	0.820	0.18	33	0.021	0.984	0.05	56	0.011	0.997	0.03
11	0.014	0.994	0.05	34	0.019	0.990	0.05	57	0.028	0.989	0.09
12	0.024	0.977	0.08	35	0.006	0.999	0.02	58	0.033	0.985	0.10
13	0.011	0.993	0.05	36	0.005	0.999	0.02	59	0.036	0.983	0.11
14	0.017	0.993	0.07	37	0.008	0.997	0.02	60	0.006	0.999	0.03
15	0.028	0.948	0.10	38	0.008	0.997	0.03	61	0.010	0.999	0.02
16	0.019	0.989	0.05	39	0.012	0.997	0.03	62	0.014	0.998	0.04
17	0.011	0.994	0.04	40	0.015	0.996	0.04	63	0.037	0.982	0.11
18	0.008	0.997	0.04	41	0.009	0.998	0.03	64	0.014	0.997	0.04
19	0.008	0.999	0.03	42	0.010	0.997	0.04	65	0.019	0.996	0.05
20	0.010	0.990	0.03	43	0.016	0.996	0.05	66	0.022	0.994	0.06
21	0.013	0.986	0.05	44	0.010	0.998	0.03	67	0.010	0.999	0.03
22	0.010	0.997	0.04	45	0.015	0.995	0.03	68	0.024	0.994	0.06
23	0.014	0.994	0.05	46	0.020	0.995	0.05	69	0.004	1.000	0.02

to verify the quality of this assertion. Figures 5.15 and 5.16 show sample results for messages on another bus, bus2, with a different configuration of the message set and a different quality of the prediction. Figure 5.15 shows one of the lowest quality results found for a high priority messages (9 out of 131), with the RMSE around 0.08. The curve representing the cdf prediction in this case is clearly separated from the simulation results and the best fitted model. However, in the case of high priority messages, although the relative error is higher, the absolute error is quite low, as the worst-case response-time is very short.

Figure 5.16 shows two sample messages (priority rank 40 and 73 out of 131) with good quality results. The method can be accurate not only for messages with one higher priority harmonic set (message priority 73), but also for messages with more than one higher priority harmonic set (message priority 40). In general, the experiments show that good quality results can be obtained. Furthermore, the errors of the prediction from statistical analysis for the reference bus and other buses are in the same magnitude. In Fig. 5.17, the comparison of the root mean squared error for each message on the reference bus and two other buses, bus2 and bus3, is given, where the X-axis is the total utilization of higher priority messages (including both local and remote loads). The average (0.0181) RMSE for messages on the *reference bus* is roughly half of the ones on bus2 (0.0322) and bus3 (0.0382), which confirms

Table 5.7 Error of the predicted distribution for messages on the reference bus with regression functions (5.9), (5.18), and (5.19)

ID	RMSE	R^2	K–S	ID	RMSE	R^2	K–S	ID	RMSE	R^2	K–S
1	0.022	0.975	0.04	24	0.011	0.997	0.05	47	0.025	0.992	0.07
2	0.019	0.980	0.03	25	0.013	0.996	0.06	48	0.010	0.998	0.04
3	0.013	0.988	0.04	26	0.010	0.996	0.03	49	0.021	0.991	0.05
4	0.013	0.987	0.03	27	0.008	0.997	0.04	50	0.013	0.997	0.03
5	0.015	0.985	0.05	28	0.012	0.995	0.05	51	0.023	0.989	0.07
6	0.016	0.983	0.05	29	0.006	0.999	0.02	52	0.026	0.988	0.07
7	0.010	0.995	0.04	30	0.013	0.996	0.04	53	0.016	0.996	0.04
8	0.010	0.996	0.03	31	0.017	0.994	0.06	54	0.030	0.985	0.09
9	0.011	0.996	0.04	32	0.011	0.988	0.03	55	0.034	0.982	0.10
10	0.013	0.986	0.05	33	0.017	0.989	0.04	56	0.010	0.998	0.03
11	0.013	0.995	0.05	34	0.017	0.991	0.04	57	0.037	0.980	0.11
12	0.006	0.998	0.03	35	0.006	0.999	0.01	58	0.043	0.974	0.12
13	0.008	0.997	0.03	36	0.006	0.999	0.02	59	0.047	0.971	0.13
14	0.017	0.993	0.07	37	0.006	0.998	0.02	60	0.007	0.999	0.03
15	0.010	0.994	0.05	38	0.008	0.998	0.03	61	0.013	0.998	0.04
16	0.005	0.999	0.02	39	0.011	0.997	0.03	62	0.018	0.996	0.05
17	0.010	0.995	0.04	40	0.020	0.992	0.05	63	0.049	0.969	0.13
18	0.012	0.994	0.05	41	0.009	0.998	0.03	64	0.020	0.995	0.06
19	0.009	0.998	0.04	42	0.010	0.997	0.04	65	0.026	0.992	0.07
20	0.008	0.994	0.03	43	0.018	0.995	0.05	66	0.030	0.990	0.08
21	0.010	0.991	0.06	44	0.010	0.998	0.04	67	0.016	0.997	0.05
22	0.008	0.998	0.04	45	0.014	0.995	0.03	68	0.032	0.989	0.08
23	0.009	0.998	0.04	46	0.022	0.993	0.06	69	0.010	0.998	0.04

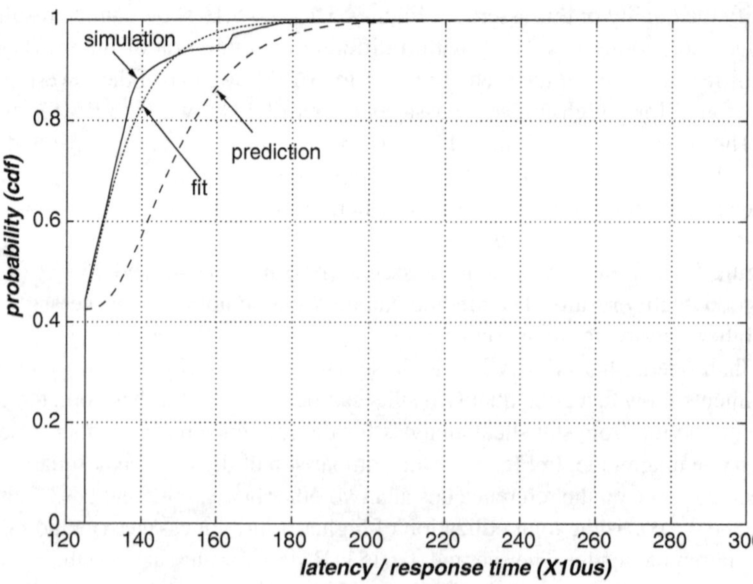

Fig. 5.15 Prediction of response-time cdf for messages on bus2: worst result

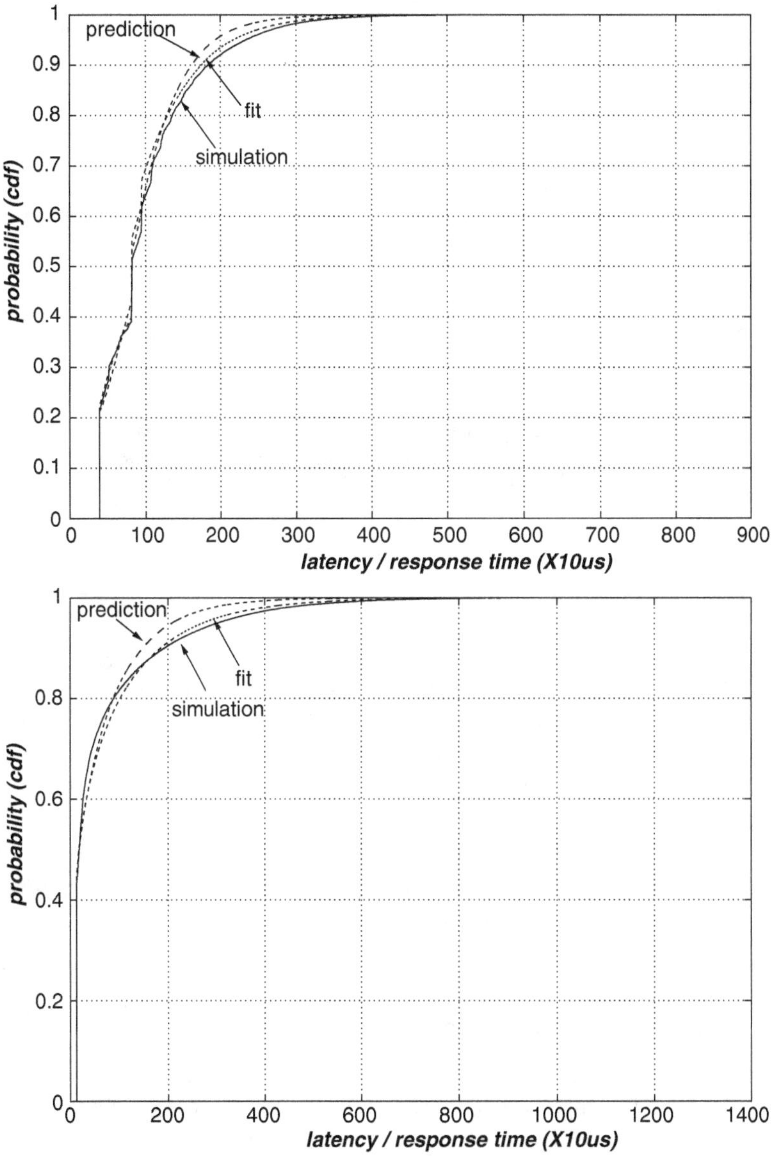

Fig. 5.16 Prediction of response-time cdfs for messages on bus2: better quality result

that the $\beta_1 - \beta_{19}$ regression coefficients estimated from the reference bus can be used to predict the response-time cdf of messages on other buses, even if at the price of a reduced accuracy.

In Fig. 5.17, the distribution of errors for messages on the reference bus looks random without noticeable pattern. However, obvious patterns can be detected in

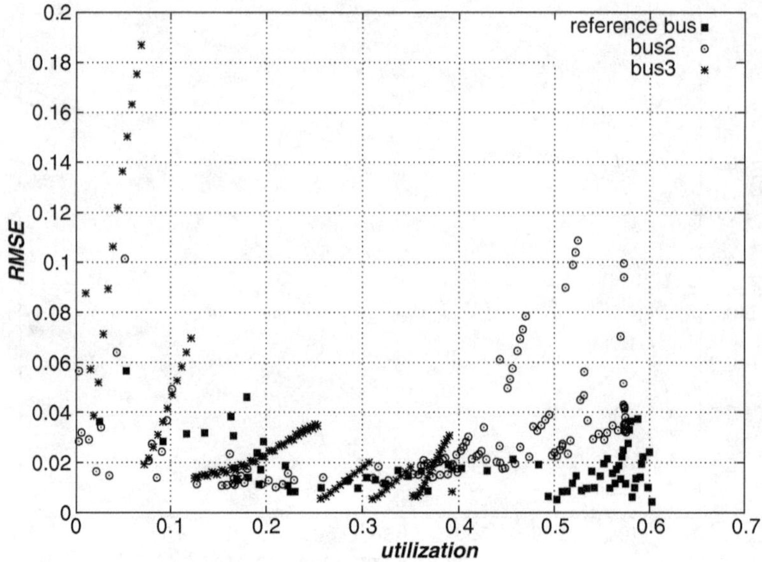

Fig. 5.17 RMSE of statistical analysis: reference bus, bus2, and bus3

the RMSE plot for messages on bus2 and bus3. For example, on bus3, when the utilization value is around 0.1, there is a segment with a linearly increasing error. In fact, the assignment of message IDs for this bus is everything but random. Blocks of messages with consecutive priority values are transmitted by the same node. Messages with relative priority 3–13 are on the same ECU, and messages with relative priority 14–24 are on another ECU. This is quite different from the reference bus. Thus, we did another experiment by modifying the message ID assignment to the reference bus according to two extreme cases:

- *ref-blk*: to assign IDs to messages on each ECU by block, i.e., ID 1–18 to ECU E_1, ID 19–34 to ECU E_2, and so on so forth;
- *ref-rand*: to assign IDs to messages randomly.

The RMSE for messages on *ref-blk* is quite high: with a maximum of 0.319 and an average of 0.0737, compared to 0.0577 and 0.0181 for the reference bus. However, the prediction for messages on *ref-rand* is much more accurate: the maximum RMSE is 0.0973 and the average RMSE is 0.0214.

Finally, Fig. 5.18 shows the results of the statistical analysis applied to messages from product vehicles, and compared not to simulation data, but to actual trace recordings. The figure shows the RMSE for the set of periodic messages (listed by higher priority utilization on the X-axis). For some messages in the set, the configuration of the node is significantly different from the one assumed by our model. First, the implementation of the communication stack, including the bus drivers, often introduces additional priority inversions and additional latencies (for

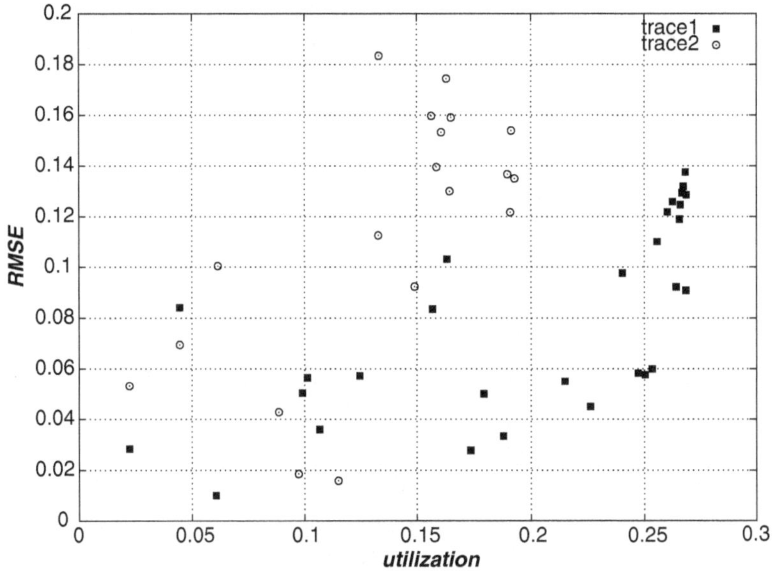

Fig. 5.18 RMSE of statistical analysis: message traces

several reasons, as discussed in Chap. 3). Second, and probably most importantly, some of the nodes are configured to transmit messages by polling (not by interrupt). These factors can further increase the error of our analysis. However, even when considering all these sources of inaccuracy, the predicted results from our analysis still remain within 0.2 in terms of RMSE, confirming its adequacy as a tool for the early evaluation of architecture configurations.

5.5 Comparison of Stochastic and Statistical Analyses

In this section, we compare and discuss the tradeoffs between statistical and stochastic analysis methods, in terms of speed of the analysis vs. the availability of data and the accuracy of results.

5.5.1 Input Information

Stochastic analysis requires complete information about the characteristics of the entire message set, including the ID (priority), source node, period, phase, and data length for each message. Statistical analysis, on the other hand, is suitable when only part of the message set is known as in the case of early design stages, such as when a new car architecture is designed and many features are partly known and subject to

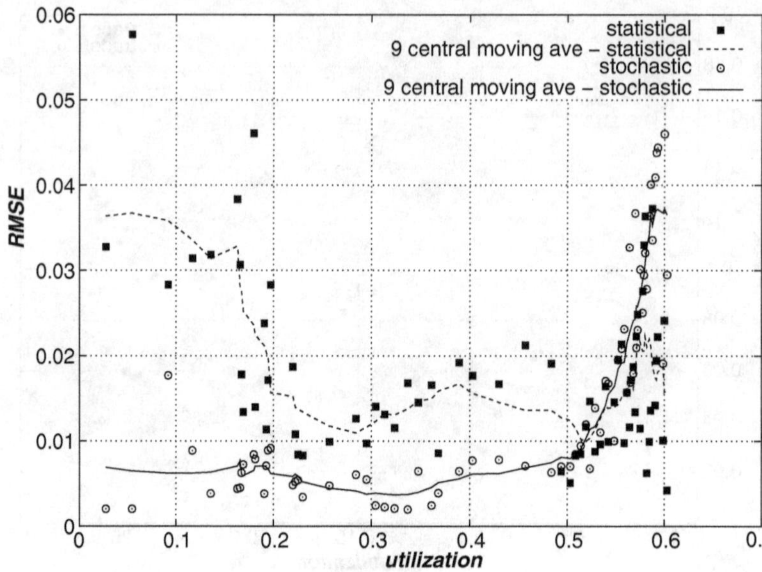

Fig. 5.19 Comparison of stochastic and statistical analyses: RMSE

uncertainty. The required information for the statistical analysis includes the local message set, the estimation of utilizations from all the messages in the system and the portion from higher priority messages, and the number of nodes in the system. Of course, if the exact local message set is unknown because of incomplete information, we can approximate it with the expected number of messages in the harmonic set, assuming that every message has maximum size, or considering an average message size.

5.5.2 Analysis Accuracy

Stochastic analysis is more accurate when complete information on the message set is available. Figure 5.19 gives the comparison of the root mean squared error and its 9-central moving average for messages in the reference bus, and Fig. 5.20 shows the comparison on K–S statistics. As in the figures, the error of stochastic analysis is almost always half of that of statistic analysis and it is mostly independent from the message priority. For high and medium-priority messages (with possible interference from a higher priority load with utilization less than 50%) the error of stochastic analysis is quite small: the RMSE is nearly always smaller than 0.01. The average RMSE for these high and medium-priority messages is 0.00577 for the stochastic analysis compared to 0.0199 for the statistical analysis. However, the error of stochastic analysis increases significantly for low priority messages and becomes even larger than the error of statistical analysis. This is because for low

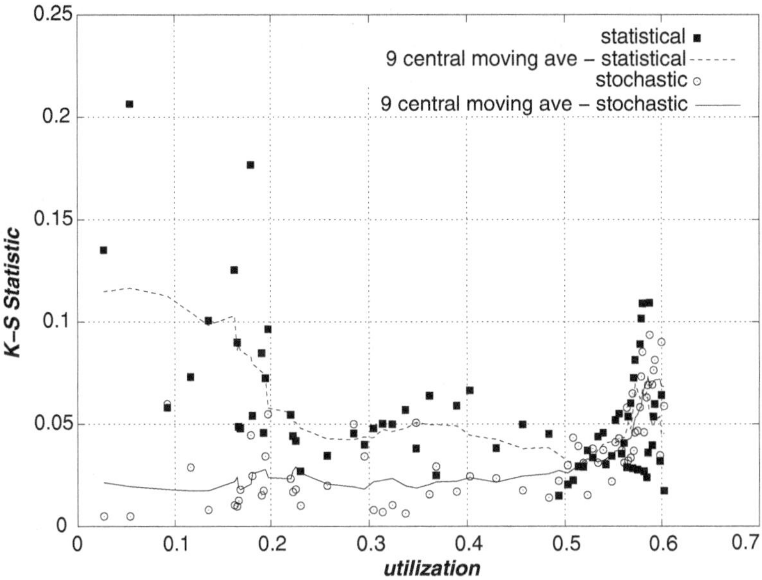

Fig. 5.20 Comparison of stochastic and statistical analyses: K–S statistics

Table 5.8 Comparison of stochastic and statistical analyses: analysis complexity

	Stochastic	Statistical
Num. of Nodes	$O(n2^n)$	$O(1)$
Num. of Messages	$O(n)$	$O(n)$
Hyperperiod	$O(n)$	$O(1)$
Worst-case response-time	$O(n^2)$	$O(1)$
Analysis granularity	$O(n^{-3})$	$O(1)$
Runtime on the example system in Table 5.1	<19 min each; 2 h total	<0.01 ms each; 0.24 ms total

priority messages, the worst-case response-times are much larger than the gcd of the message periods, i.e., the period of the TxTask. These messages have a significant probability of experiencing multiple interferences from higher priority messages and the correlation between the message queueings of the messages providing interference is not captured with accuracy by our stochastic analysis model [65]. On the contrary, statistical analysis is more robust to the occurrence of multiple interferences from high rate periodic messages.

5.5.3 Analysis Complexity

Stochastic analysis is more accurate when the complete message set is available. However, it is significantly slower, as shown in Table 5.8. The complexity of stochastic analysis is very sensitive to many system parameters. For example, the complexity is worse than exponential with respect to the number of nodes in the system, and increases with the cube of the number of time ticks (reciprocal to the time granularity of the analysis) in the least common multiple of the message periods. However, statistical analysis is independent from almost all system parameters, since it involves only the evaluation of several closed-form formulas. The runtime information[2] on the example system as in Table 5.1 is also compared, which indicates that statistical analysis is more than 10^7 times faster than stochastic analysis.

[2]Java implementation on laptop with 1.6 GHz CPU and 1.5 G RAM.

Chapter 6
Reliability Analysis of CAN

In this chapter, we will discuss the reliability of CAN communication systems. Because the subject of reliability may not be familiar to all readers, we first provide some general definitions that are going to be used in the rest of the chapter. *Reliability* is the ability of a system to perform its specified function for a prescribed time and under stipulated environmental conditions. Reliability, along with availability, safety, and other metrics, is part of the more general concept of dependability.

In contrast, a *failure* is a situation in which a system is not performing its intended function. *Faults* are defects in a part of a system that may or may not lead to failures as, although a system may contain a fault, its input and state conditions may never cause this fault to be executed so that an error occurs and thus never exhibits as a failure. An *error* is a manifestation of a fault as an unexpected behavior within the system. Errors occur at runtime when some part of the system enters an unexpected state due to the activation of a fault. Errors may (or may not) lead to failure. They may not if the system has some fault tolerant mechanisms that can catch the fault before it becomes an error and subsequently a failure. If faults can propagate into system-level failures, *hazards* may occur that have the potential to threaten injury or loss of life.

Generally speaking, the reliability of the CAN protocol against transmission errors is a topic of paramount importance that any textbook should address as it is related to assessing the robustness of the protocol to faults and its potential to create hazardous situations. In the description of the protocol standard in Chap. 1, we presented the basic features for atomic broadcast, error signalling and general management, including error detection and confinement (by the isolation of faulty components) and error recovery. In this chapter, we touch on several additional issues that relate to the general topic of the protocol reliability—we try to quantify the impact of different errors, and discuss scenarios that could lead to inconsistent duplicates or omissions. We also address "built-in" protocol vulnerability to errors that could be relevant in evaluating the use of CAN in high-reliability applications.

M. Di Natale et al., *Understanding and Using the Controller Area Network Communication Protocol: Theory and Practice*, DOI 10.1007/978-1-4614-0314-2_6,

6.1 Error Rates in CAN

Several research papers have been published discussing the potential impact of CAN faults on the message timing (especially their worst case latencies) and the overall reliability of the network. In almost all of them, the quantitative evaluation of the probability of occurrence of these faults and their practical impact requires the definition of an error model. This probability is measured as a function of a parametric error rate (at the bit- or frame-level). Of course, the applicability of the results depends on the actual numbers that are expected for the bit error rate (*ber*) or the frame error rate. Error rates, in turn, depend on the bus physical characteristics and the amount of Electro-Magnetic Interference (EMI) that is expected in the working environment.

Unfortunately, there is very little real data available on the experimental evaluation of the bit error rate for a CAN bus, with two exceptions. [45] presents a study on a vehicle in which the amount of error caused by EMI is qualitatively estimated based on the number of message retransmissions and the corresponding message delays because of errors (the actual bit or frame error rate is not provided). According to the author's conclusions, CAN is shown to be a robust communication protocol in the harsh, real-world vehicle environment with significant electromagnetic activity. In the presented experiments, the bus was able to recover quickly from all errors, with no loss of data or significant delay.

The only publicly available work in which bit error rates are explicitly measured is [26]. Indeed, the difficulty in the evaluation of this basic reliability parameter is quite justified, since it is not easy to measure the actual bit error rate using a standard CAN controller. The only information directly provided by the controller is the occurrence of a frame error.

The measurement system developed and described in [26] is represented in Fig. 6.1. It consists of a board with a programmable HW, in which a MoiCAN IP core is the message receiver, implementing a subset of a CAN 2.0A controller, and optimized for deployment onto a Xilinx FPGA. The transmitter is a board based on

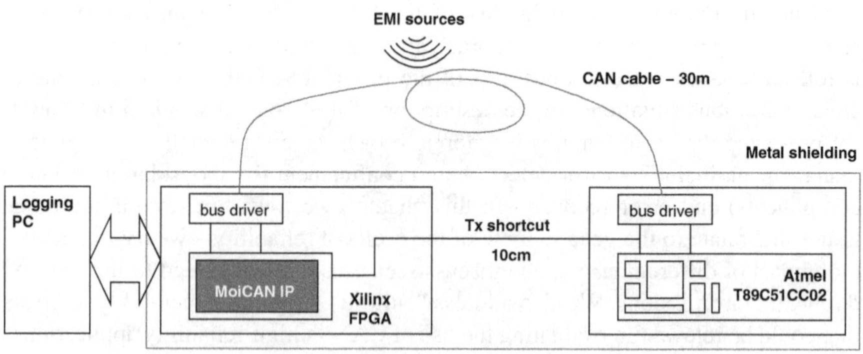

Fig. 6.1 The experimental setup for measuring bit error rates in [26]

Table 6.1 Bit error rates on a CAN bus measured in three different environments

	Benign environment	Normal environment	Aggressive environment
Bits transmitted	2.02×10^{11}	1.98×10^{11}	9.79×10^{10}
Bits errors	6	609	25,239
Bits error rate	3×10^{-11}	3.1×10^{-9}	2.6×10^{-7}

the Atmel T89C51CC02 CAN controller. Finally, a log recording PC is connected to the FPGA board with the MoiCAN controller through a parallel port. The bus transceiver and driver used on both the transmitter and receiver side is the Philips 82C250.

Both controllers are configured for transmitting at 1 Mbps. The physical connection is a 30 m cable (unfortunately not otherwise specified). The bitstream is also transmitted using a 10 cm internal shortcut to the MoiCAN board, where the two streams are bitwise compared. Besides the standard CAN controller logic, the MoiCAN implementation in the programmable HW also hosts a dedicated register that is used to count the number of bit errors detected when comparing the two streams. The Atmel controller transmits 8-byte data frames every 400 μs, using approximately 25% of CAN bus bandwidth.

Three sets of experiments were conducted: one at the laboratories of the university (marked as *benign environment*), another at a factory, two meters away from a high-frequency arc-welding machine (*aggressive environment*), and finally the third at a factory production line (or *normal environment*). The CAN cable was folded in order to make a winding with just one turn and the MoiCAN and Atmel boards were shielded within a metal case. The measured error rates obtained at the end of the experiments are shown in Table 6.1.

The authors of the study note that a preliminary analysis of the temporal distribution of the errors indicates that in the aggressive environment (where most errors are detected) around 90% of the errors were part of bursts with an average duration of 5 μs.

These numbers cannot be used as an accurate reference when discussing the reliability of CAN, given the unknown type of connecting cable, the influence of the type and strength of the electromagnetic interference in the specific case (for which no quantitative measures were made available), and possibly other factors. However, they could provide a first-order estimate about the bit error rates that could be expected for a CAN-based communication. In the rest of this chapter we discuss issues for which the estimated impact may be significant for bit error rates of 10^{-4} and much less for bit error rates around 10^{-7}. The experimental results in [26] will be used to discuss the impact of the possible faults and put in a more realistic perspective of the issues that will be highlighted in the rest of the chapter.

In detail, we will discuss the occurrence of failures, which include inconsistent message omissions and duplicates, and the unintended validation of corrupted messages. All these events have been detected as possible, but characterized by a very low probability, especially if evaluated using the bit error rates shown in Table 6.1.

However, a careful quantitative evaluation of the impact of very low probability events is often necessary, given the use of the CAN bus in car industry. For example, the U.S. automotive fleet logs approximately four orders of magnitude more operating hours annually than the U.S. commercial aviation fleet. Thus, a rate of 10^{-9} per hour that results in a failure event every 73 years in the commercial aeronautics fleet, will produce a failure every 4.5 days in the much larger automotive fleet [37].

Before discussing these issues, we shortly touch upon some deviation points in the standard. While they might not necessarily appear to be reliability issues, the handling of unexpected message content/format needs to be clarified to ensure the correct process of messages and avoid inconsistent behaviors, such as inconsistent message omissions.

6.2 Deviation Points in the Standard

Unfortunately, several published CAN specifications and standards contain errors or are incomplete. The incomplete definition of the physical layer in the original Bosch specification is one example of such omissions. To avoid CAN implementations that are de-facto incompatible, that is, to prevent the situation in which two HW controller implementations are unable to communicate with each other, Bosch made sure (and still does) that all CAN chips comply with the Bosch CAN reference model. Furthermore, the University of Applied Science in Braunschweig/Wolfenbüttel, Germany, has been conducting CAN conformity testing for several years, led by Prof. Lawrenz. The test patterns in use are based on the internationally standardized test specification ISO 16845.

Gaps in the specifications have been filled in incrementally. When ISO took charge of CAN specifications, several revisions have attempted at fixing omissions and errors. Unfortunately, quite a few still remain and some of them are unavoidable, because fixing them would require reworking some of the protocol features completely. In this section we discuss some of these issues.

6.2.1 Incorrect Values in the DLC Field

The first problem is with the Data Length Code (DLC) field. According to the original CAN specification by Bosch, no transmitter may send a frame with a DLC value larger than 8. However, the case of a message in which accidentally the DLC field is set to a value greater than 8 is not covered by any of the error types defined by the standard. This is indeed neither a Bit Error, nor a Stuff Error, nor a CRC Error, nor an Acknowledge Error. It could be regarded as a Form Error, but the DLC belongs to the stuffed Control Field and the Form Error is only defined for the fixed-form bit fields. In the end, while this is clearly an out-of-specification condition, there is no error or error signalling defined for it and the receiver behavior is not specified. Obviously, this situation could lead to a failure as the error is not caught at its source. In this case, the Reference CAN Model provides the following recommendation: any receiver that receives a frame with a DLC field value greater than 8 should expect to receive 8 bytes (that is, if the received DLC > 8 then DLC $= 8$).

6.2.2 Dominant SRR Bit Value

The SRR bit is another possible source of misinterpretation. The original CAN specification requires the SRR bit to be sent as recessive, however, the behavior of the receiver in case of a dominant SRR bit is not specified. Once again, this is clearly not a Bit, Stuff, CRC or Acknowledge Error. In addition, since the SRR bit is located in a stuffed bit field, it is not a Form Error. The Reference CAN Model assumes for the SRR bit a behavior similar to the reserved bits, which have to be sent as dominant, but whose actual value is ignored by receivers. So no transmitter may send a dominant SRR bit in an Extended Frame while a receiver ignores the value of the SRR bit (but the value is not ignored for bit stuffing and arbitration). Since the SRR bit is received before the IDE bit, a receiver cannot decide instantly whether it receives an RTR or an SRR bit. That means only the IDE bit decides whether the frame is a Standard Frame or an Extended Frame.

6.3 Fault Confinement and Transition to Bus Off

CAN error management aims at identifying errors that are caused by short disturbances and are temporary or, at worst, intermittent as opposed to permanent hardware failures. In the first case, error counters can at most bring the node in an error passive state, while the bus off state should only be reserved for permanent error conditions.

The probability of a network controller to be in one of the degraded modes (bus passive or bus off) has been analyzed in [30] using Markovian analysis based on a parametric bit error rate. The idea is to evaluate the probability that a node enters a bus passive mode (in which the time performance of its outgoing messages is expected to be degraded) or even bus off state, simply because of bursts of bit errors which depend on external causes (such as EMI) and not because of the faulty state of the node itself (as implied by the CAN specification).

The details of the continuous time Markov model (and its discrete time approximation) can be found in the referred paper. Here, we only discuss the final results, based on the graphs shown in [30]. The average hit time of the bus off state depends on the message load and the bit error rate. For the two cases taken into consideration in the paper (an engine controller with a relatively low network traffic and a network gateway with higher load), the average hit time is 1×10^{15} h for the engine controller and 1×10^{12} h for the gateway, for an estimated *ber* of 5×10^{-4}. Furthermore, these times are rapidly decreasing with lower bit error rates (for example, the gateway average bus-off hit time ranges from 1×10^{8} to 1×10^{12} h for a *ber* going from 6×10^{-4} to 5×10^{-4}. If these expected times are computed using the bit error rates experimentally measured in [26], the resulting values are so low to dismiss the scenario that a node enters the bus-off state because of external interferences in practically all cases of interest.

6.4 Inconsistent Omissions or Duplicate Messages

CAN was designed to allow atomic broadcasts except for consistent omissions and node failures, that is, a message stream is either totally delivered to all nodes or to none of them. Unfortunately, this is not true and inconsistency scenarios may happen (with a quite low probability as discussed next). Bit errors in the last two bits of the end-of-frame delimiter can cause inconsistent message delivery and generation of inconsistent duplicate messages or inconsistent omissions.

The problem was first highlighted and discussed in [34] and stems from the following protocol feature. In CAN, errors are signalled only up to the last bit of the interframe sequence. If the last interframe bit is incorrectly received (that is, if its value is dominant instead of recessive), nodes are required to ignore the error (the intuition is that there would be no way to detect the following error frame, given that the frame terminates anyway and the error frame could not be distinguished from an overload frame). Therefore, if the sender detects no errors up to the last bit, it considers the transmission successful and does not retransmit.

However, if a disturbance corrupts the last but one bit in the interframe sequence for only a subset of the receiver nodes, denoted as **e**, all nodes in **e** will start signalling the error in the last interframe bit and will discard the message. The sender itself will detect the error and initiate a retransmission. However, the receivers not in **e** (denoted as $\bar{\mathbf{e}}$ for convenience) will accept the message frame since they only detected an incorrect value in the last interframe bit and the standard requires that they ignore it. Also, the following error frame will be interpreted by these receivers as an overload frame. When the sender retransmits the message, these nodes will have an inconsistent message duplicate (or IMD). Even worse, if the sender node crashes before the retransmission, the nodes in **e** will have an inconsistent message omission (or IMO, Fig. 6.2).

In the Time-Triggered CAN protocol (or TT-CAN, see Chap. 9), because of the time-triggered nature of the transmissions, nodes do not automatically retransmit in the case of a fault (the network controller must be put in the so-called one-shot mode). In this case, any occurrence of the previous scenario will result in an inconsistent message omission.

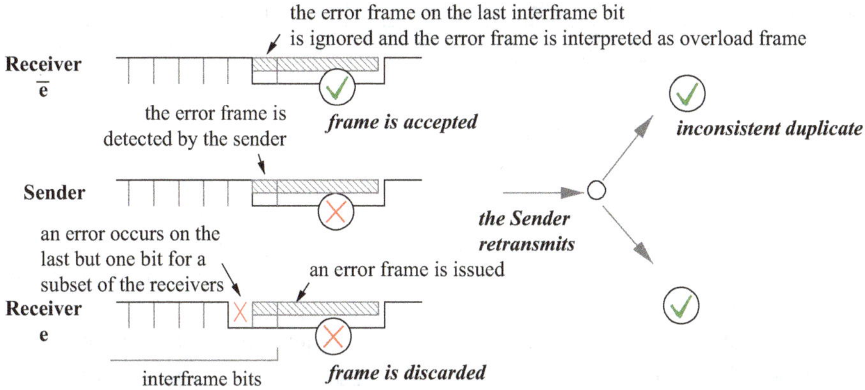

Fig. 6.2 A scenario leading to inconsistent message duplicates

Given a value of the bit error rate or *ber*, the probability of inconsistent frame omission can be evaluated as

$$p_{ifo} = (1 - p_{ced}) \times (1 - ber)^{L_{msg}-2} \times ber$$

where p_{ced} is the probability of a consistent error detection by all nodes in the network, assumed as 0 in [34] (which leads to a discordance with the experimental results later found in [26]); L_{msg} is the message frame length (in bits). The formula is based on the assumption that the probability of having an error in a given bit position of a frame is defined by a geometric distribution, because it is assumed that the sender stops transmitting after the detection of the first bit error. Again, this implies that the sender always detects the bit error. If a bit error is only detected by the receiver, then the error frame will only be sent if it causes a frame error or a form error.

The probability of node crash failures between the end of a transmission and the beginning of a new one is computed as follows:

$$p_{fail} = 1 - e^{-\lambda \times \Delta t}$$

assuming Δt is the time interval between the end of a transmission and the end of the next one, and λ is the failure rate of network nodes. Finally, given the previous, the probability of inconsistent message duplicates (IMD) is computed as follows:

$$p_{IMD} = p_{ifo} \times (1 - p_{fail})$$

and the probability of inconsistent message omissions (IMO) is

$$p_{IMD} = p_{ifo} \times p_{fail}$$

Table 6.2 Probability of inconsistent message omissions and inconsistent message duplicates

	Benign environment	Normal environment	Aggressive environment
Bit error rate	3×10^{-11}	3.1×10^{-9}	2.6×10^{-7}
IMO/h (CAN)	$p_{ied} \times 1.22 \times 10^{-13}$	$p_{ied} \times 1.24 \times 10^{-11}$	$p_{ied} \times 1.05 \times 10^{-9}$
IMD/h (CAN), IMO/h (TTCAN)	$p_{ied} \times 8.75 \times 10^{-4}$	$p_{ied} \times 8.93 \times 10^{-2}$	$p_{ied} \times 7.59$

Table 6.2 shows the probability of inconsistent message omission based on the bit error rates measured in [26] and the probability of an error being detected by only a subset of the receivers, labeled as $p_{ied} = (1 - p_{ced})$.

The probability of inconsistent message duplicates has been experimentally measured in [26] to be equal to 1.8×10^{-2} in the aggressive environment experiment. This would imply that (at least in that experimental setting), the probability that a bit error is detected only by a subset of the receivers p_{ied} is quite low, in the order of 2×10^{-3}.

Overall, these events appear to be quite unlikely to occur, but in some applications inconsistent message duplicates or even omissions could pose a problem even at its very low probability of occurrence, and need to be mitigated or even nullified by application-level protocols that detect such inconsistencies. We do not discuss the details of these protocols further. The interested reader should refer to [34] for more details.

6.5 Protocol Vulnerability

The scenario in which network errors are undetected is one of the most critical types of failures in networks. Error detecting codes and redundancy checks (CRCs) can detect cases in which the message content is corrupted in a subset of its bits. CRCs are purposely designed to detect errors occurring with a given probability so that application designers can assume that a message validated by the CRC check is indeed correct.

Of course, there may be cases in which the assumptions on the effectiveness of the error detection are not correct. Unfortunately, because of interactions between the bit stuffing mechanism and the CRC protection domain, the CAN protocol exhibits vulnerabilities to multi-bit errors, so that the CAN protocol can accept a small fraction of messages with as few as two bit errors and, of course, even a larger number of bit errors with increasing probability.

The CAN specification states that the CRC field will detect all burst errors up to 15 bits, and all errors with 5 or fewer disturbed bits. It also states that other multi-bit errors (with at least 6 disturbed bits or burst errors of more than 15 bits) will slip through undetected with a probability of 3×10^{-5}. The problem has been first described in [61]. In presence of only two bit errors, the bit stuffing protocol can modify substantially the sequence of transmitted bits, to the point of creating errors (not necessarily bursty) quite long, easily longer than 5 bits, that can possibly go undetected by the CRC.

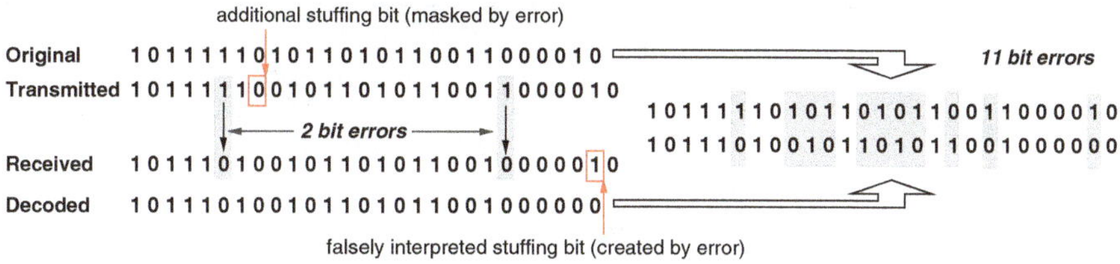

Fig. 6.3 Two bit errors result in a much longer sequence because of stuffing

Figure 6.3 shows a scenario with a possible missed detection in a message in which only two bits are flipped (change value) because of, for example, EMI interference. In this scenario, the transmitted stream includes an early sequence of five bits at one, followed by a stuffing bit. During transmission, a bit error changes the value of one of the five ones in the sequence. The receiver will therefore assume there was no stuffing and reads the entire sequence, including the bit at zero that was added as stuffing, now considered as a regular data bit. If this were the only error, the sequence would eventually result in an illegal message length, or another type of message form error. However, in our example, a second bit error (a bit with value one turned into a zero late in the sequence) creates a sequence of five zeros that was originally not in the transmitted stream. Hence, the receiver interprets the following bit at 1 as a stuffing bit (incorrectly), therefore compensating the error in the message length. Of course, the bit errors might as well result in an additional early stuffing bit and a late omitted one. In both cases, when comparing the two sequences, these two bit errors result in a shifting of the entire bit sequence with a possibly large number of errors, so large that it can slip through the error detection capability of the CRC code. In our example, there are 11 bit errors between the transmitted and the received stream. The errors are not even necessarily in the form of a burst, given that the sequence shift can generate an almost arbitrary combination of bit errors.

Unfortunately, as stated in [61], it is very difficult (for a number of reasons, including the variable length of messages and the bit stuffing mechanisms) to define an analytical approach that predicts the probability of occurrence of this problem given the bit error probability. Therefore, a simulation was performed to experimentally evaluate the probability of this type of errors.

The simulation program creates a random sequence of CAN messages in which bits are corrupted during the simulated network transmission stage, by randomly changing the value of a single bit (flipped bit) or forcing a burst error (a sequence of variable length of bits of the same value). For randomly flipped bits, the simulation parameter is the number of bit flips, and the simulator randomly defines their positions. For burst errors, the parameter is the burst length, and the simulator randomly defines the position of the starting bit of the burst. The simulator generates messages with random ID and data content.

Upon reception, the receiver performs the de-stuffing (when and if needed) and all the regular CAN message checks (form errors, length errors, and bit field

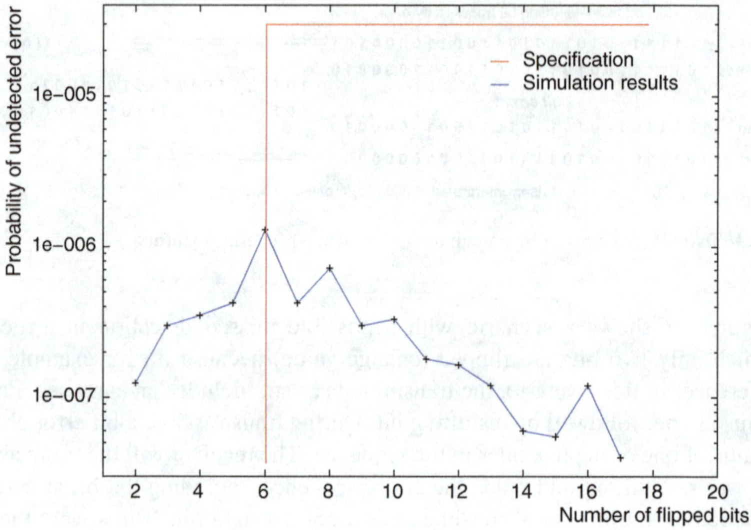

Fig. 6.4 Probability of a corrupted message being undetected for a given number of bit errors

errors). The receiver also checks for illegal stuffing patterns (i.e., six or more zeros or ones in a row) and other message format violations. In the majority of the cases, the corrupted message will be dropped due to either stuff or form errors. But in a small number of cases, the message may pass both the stuff and form error checks and a new CRC is calculated from the corrupted message. If the computed and transmitted CRCs are the same, despite the injected errors, then a failure occurs in the form of a corrupted message being accepted as valid.

All original messages in the simulation used 8-byte data fields and 11-bit ID fields. All simulations were run to ensure at least 100 failures, and more than 1,000 such failures in selected instances. A two-state symmetric binary channel model was used for the CAN physical medium where the probability for a bit to change from dominant to recessive is the same as the probability for a change from recessive to dominant. The results are shown in Fig. 6.4 by plotting the probability of undetected errors as a function of the number of bit errors, along with the probability as prescribed by the CAN CRC specification (3×10^{-5} for 6 or more bit errors). The number of corrupted messages that are accepted after the CRC checks first increases, from 2 to 6 bit errors, and then it decreases with an increasing number of corrupted bits (contrary to what one might expect, and as implied by the CAN specification) because of the increased probability that the message will cause a stuff error or form error. Clearly, the end result is an unexpected vulnerability to errors of 2 to 5 bits, albeit with a low probability.

It is at this point interesting to evaluate what could be practical impact of this unexpected source of failures. The probability that a message with 2 flipped bits is undetected and accepted is 1.29×10^{-7}. There are probably approximately two hundred million ground vehicles in the U.S. and 2.469 trillion vehicle miles [37]

Table 6.3 Number of undetected message errors per day as a function of the bit error rate

	Benign environment	Normal environment	Aggressive environment
Bits error rate	3×10^{-11}	3.1×10^{-9}	2.6×10^{-7}
UME/day	4.5×10^{-11}	4.5×10^{-7}	3.4×10^{-3}

traveled per year at an average speed of approximately 30 miles per hour, which results in an accumulated operating period of 2.96×10^{14} s. Also, assuming that there is one CAN bus in each vehicle operating at 50% utilization, or 500 kbps total utilization, and that each message is 117 bits in length including bit stuffing, there is a total of 1.48×10^{20} bits or 1.27×10^{18} messages transmitted per year and subject to corruption.

Due to lack of better estimates, the computations performed in [61] for the probability of undetected error faults assume that the CAN *ber* is the "typical" error rate of a copper cable, estimated to be around 1×10^{-6} [38] and possibly higher because of the noisy automotive environments in which CAN buses typically operate. For a bit error rate of 10^{-6}, the expected number of undetected message errors in the entire US fleet of vehicles would be 0.05 messages per day. However, this number would decrease with the square of the *ber*. Table 6.3 shows the expected number of undetected errors (UME) using the bit error rate measurements in [26], which are much lower than the estimates in [61].

In applications where this vulnerability is still considered a risk, there are ways to improve the error detection capability by using one or two bytes in the data content of the message for an additional CRC or checksum. Using two bytes to compute an additional 16 bit CRC, all corrupted messages were detected by the CAN simulation. However, an additional single-byte CRC only results in a very slight decrease in the number of accepted corrupted messages. A one-byte checksum performs much better, improving the error detection by 2 orders of magnitude. A Longitudinal Redundancy Code (LRC) is also presented in [61] because it has a cheaper hardware implementation than an adder. The 8-bit LRC has a better error detection capability than the 8-bit CRC, but not as good as the 8-bit checksum.

6.6 Bus Topology Issues

Another source of vulnerability of CAN networks is at the physical level. The most common bus configuration for CAN systems is still a single shared bus line, which is clearly a single point of failure for the system. Physical malfunctionings can occur at different devices. The bus line may be truncated or disconnected, or shorted, or grounded; connectors may come loose, transceivers may be faulty. In all these cases, the network may end up being partitioned, with each partition working to some degree or not working at all. Replicated buses may alleviate the problem, but they hardly constitute a final solution, given that in most cases, the replicated buses must join in the proximity of the nodes.

The solution, as for other network standards, including Ethernet, FlexRay and TTP (the Time-Triggered Protocol) consists in a star topology, in which the star concentrator takes care of automatically disconnecting the bus connections that are malfunctioning, isolating the faulty links and nodes. Star concentrators of this type have been proposed in the CANcentrate (and ReCANcentrate) projects [15,56], but no such device has been produced commercially until now.

6.7 Babbling Idiot Faults and Bus Guardians

Besides physical failures, what is probably the most serious concern in a CAN network is the event of a node behaving as a *babbling idiot*. In this case, a node may (permanently or sporadically) erroneously transmit a message stream (possibly with a low identifier) with a higher rate than its design specification, or maybe simply produce messages earlier than expected. In this case, nothing prevents a (faulty) flow of high priority messages from disturbing the communication of other nodes, to the point of possibly bringing them to starvation. The standard CAN protocol has no solution to this problem, which needs of course an appropriate treatment in safety-critical systems. The solution should consist in the detection of nodes that are misbehaving and their isolation from the network. Of course, the detecting device must be external to the transmitting node and capable of disconnecting it from the network. Such device is conventionally called a *Bus guardian*.

Bus guardians for CAN have been discussed and proposed in the scientific literature [17, 31, 35], both as devices working in pair with the transmitting nodes, and as devices integrated in a start concentrator for star topologies. Commercial bus guardians are however not available today, quite possibly because the reliability of simple CAN networks has been in practice quite sufficient for handling today's applications and the use of CAN for high-reliability, high-integrity systems is still not quite seen as a near-future option.

Chapter 7
Analysis of CAN Message Traces

The previous chapters discuss different analysis techniques for computing the (worst-case and average-case) response time of CAN messages. These methods are based on a number of assumptions, including the availability of a perfect priority-based queue at each node for the outgoing messages, the availability of a transmit object, TxObject, for each message, the preemptability of the transmit objects, and the ability to immediately (i.e., in zero time) copy the highest priority message from the queue to the TxObjects at the source node as soon as they are available. When these assumptions do not hold, as in many cases of industrial relevance, the response time of messages can be significantly larger than what is predicted.

In many practical cases, the system under study does not satisfy the conditions for the analysis techniques discussed in the previous chapters. In fact, in some cases, the complete system configuration may not be available to the analysts.

In these cases, the starting point for analyzing the timing characteristics of the system is the set of message traces, extracted with tools such as CANalyzer from Vector [11]. A message trace contains the instants at which each message is transmitted (with the corresponding identifier and content) and a possible evidence of a timing fault. This chapter summarizes methods that can be used to reconstruct the following from the message traces:

- An estimate of the message queuing times.
- The reconstruction of the possible message arrival times starting from their receive time stamps.
- The analysis of the hardware clock drift at each node.
- The detection of true message periods and grouping of messages by transmission node (if not available).
- The detection of the queuing and transmission policies used at the middleware- and driver-level.
- The probability distribution of the response times of the queuing tasks (the message queuing jitter).

M. Di Natale et al., *Understanding and Using the Controller Area Network Communication Protocol: Theory and Practice*, DOI 10.1007/978-1-4614-0314-2_7,

From the above estimates, details about the (possibly unknown) system can be derived. We will discuss two case studies to show practical applications of the techniques. These examples demonstrate how trace analysis can detect the source of timing anomalies, as well as help reconstruct the message-to-ECU relation in a bus for which this information is not available.

7.1 Notation

For the purpose of timing analysis, each periodic message m_i is characterized by the tuple $m_i = \{b_i, id_i, \Upsilon_i^{\text{src}}, T_i, J_i, \phi_i\}$, where b_i is the length in bits (multiple of 8), id_i is the message identifier, Υ_i^{src} its sending node, T_i its period, J_i its queuing jitter. Messages are transmitted according to their identifiers (priorities). For each periodic activation, we consider a message instance. We denote the j-th instance of message m_i as $M_{i,j}$, its *arrival time* as $a_{i,j} = \phi_i + (j-1) \times T_i$, which is the time $M_{i,j}$ is logically ready for execution. B_r is the bit rate of the bus and $p = 46$ is the number of protocol bits in a standard frame. The message length needs to account for bit stuffing. $c_{i,j} = \frac{b_i + p + n_s}{B_r}$ is the time required to transmit the message instance provided that it wins the contention on the bus, where n_s is the number of stuff-bits.

The middleware-level TxTask is activated periodically and has the responsibility of assembling each message from the signals. The period of each message m_i is an integer multiple of the TxTask period T_{Tx} ($T_i = k_i T_{Tx}$). In case TxTask always enqueues messages at the end of its execution, the activation jitter of message m_i can be computed as the worst-case response time of TxTask (or better, the difference between its worst- and best-case response times).

The trace $\mathbb{T}$ consists of a finite sequence of events e_i, $\mathbb{T} = \{e_0, e_1, \ldots, e_n\}$, where each event is a tuple $e_i = \{t_i, id_i, \mathbf{v}_i\}$: t_i is the event time stamp, id_i is the CAN id of the message to which the event is referred and $\mathbf{v}_i$ is a vector of bits, which defines the data content and the corresponding number n_s of stuff-bits.

7.2 Case Studies

The first case study consists of an automotive system in which a timing fault is detected and the system needs to be analyzed for understanding the cause of the error and to provide possible remedies. The fault is an overwrite error for a message transmitted periodically and received by a polling task. Because of excessive delays, the separation between the receive times of two message instances falls below the safe margin that ensures reading at the destination before overwriting. The system consists of a vehicle CAN bus with five nodes (E_{11} to E_{15}), and a tracing node, which records the receive time and the content of all the messages that are transmitted on the bus. The message with identifier **0x1F3** from node E_{11} is transmitted with a period of 20 ms, and some of its instances arrive at the destination

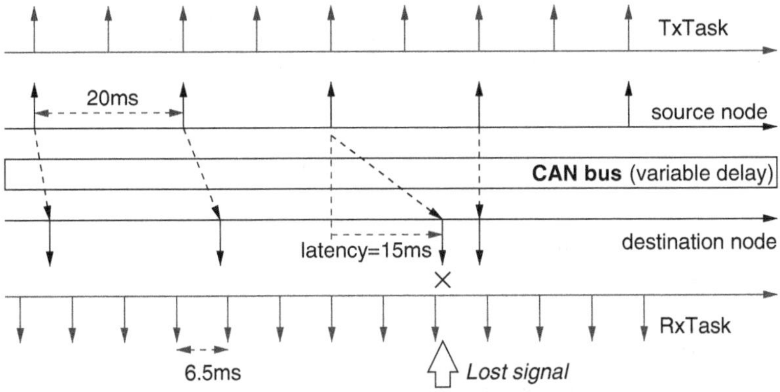

Fig. 7.1 Signal overwrite error

node with a separation of less than 5 ms, therefore with a worst case response time larger than 15 ms. At the receiving node, a polling task checks the incoming CAN registers with a period of 6.25 ms, and misses instances of **0x1F3** because of overwriting, hence causing an error (Fig. 7.1). A message response time of more than 15 ms is extremely high, exceeding the worst-case value predicted by the analysis.

The second case study consists of another automotive bus (from a different system), in which the supplier of a connected node wishes to perform worst-case timing analysis for some of its outgoing messages. However, for IP protection, the system integrator is not providing the configuration of all the messages transmitted on the bus by the other nodes. Therefore, in order to recover the information that is necessary for the analysis, the only source of information are the (long) message traces recorded on the bus.

7.3 Trace Analysis

The starting point for the analysis is the message trace data archived by the tracing node. Each line of the trace identifies a message transmission event. Each event is labeled with the time instant at which the message has been received by the tracing node (i.e., the time stamp of the event), the CAN id, the data length (in bytes) and the actual data content. In addition, when available, information about the source node and the expected period for all the messages on the bus may also be used. In this case, each message is identified by the CAN id, its expected period, and its source node.

The first step of the analysis is to reconstruct the arrival times of messages based on their reception times. The analysis is performed by reasoning backwards in time,

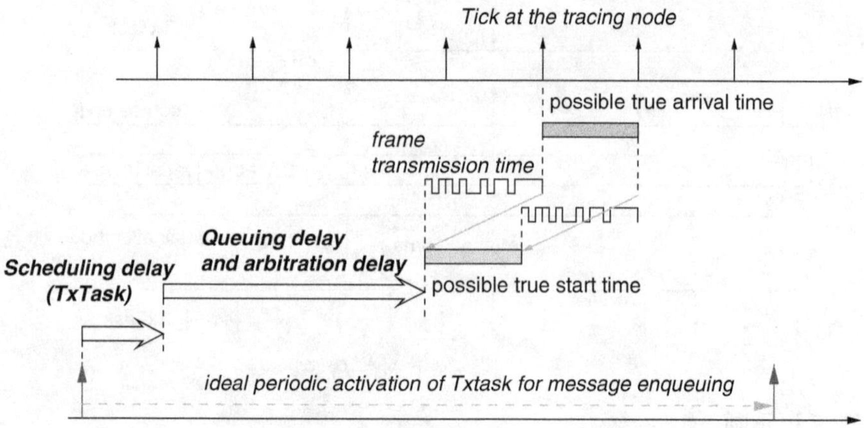

Fig. 7.2 From message arrival to message receive

and subtracting the factors that contribute to the message response time (Fig. 7.2). The message reception time is recorded with a finite resolution *res*, which depends on the time stamping clock at the tracing node. This resolution identifies a window of possible reception times $[t_i, t_i + res)$ that would result in the recorded time stamp. By subtracting the message transmission time, the window of the possible starting time for the message transmission is obtained. The message transmission time can be obtained from the trace. The number of stuff-bits can be computed from the length of the message data and the actual data content of the message. The following are still not accounted for: the *queuing and arbitration delays* of the message, and the *scheduling delay* of the TxTask which is responsible for adding the message to the local transmission queue. The message queuing and arbitration delays are unknown, except for specific message instances in the trace, where these delays can be assumed to be very short and negligible with respect to the other time quantities. Special queuing events corresponding to message instances with negligible queuing delay can be identified in the trace since they follow a time interval in which the bus is idle. Identification of such instances is therefore our next step.

7.3.1 Identification of Reference Messages

For each message in the trace, we compute the window of the respective receive and start transmission times. When the start transmission time window does not overlap with the receive time window of the predecessor (i.e. their intersection is empty, see Fig. 7.3), a bus idle time interval between the two messages is identified and the message that is transmitted after the idle time is labeled as a *reference* instance with zero queuing and arbitration delay.

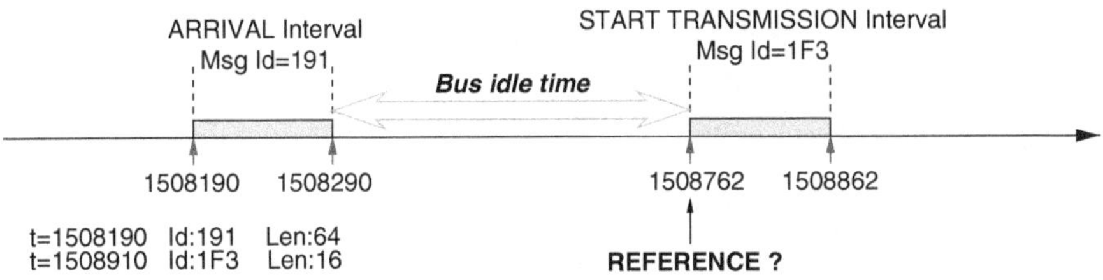

Fig. 7.3 Detection of idle time on the CAN bus

As shown in the following, the idle time interval should be at least larger than the interrupt handler response time to ensure that the following message did not wait in the queue. However, the existence of any idle time is a sufficient proof to consider the arbitration delay to be zero. Idle time for the detection of reference instances is in practice always found in automotive buses, which seldom (if ever) exceed 70% utilization for several reasons, including tolerance with respect to message loss, extensibility and the need to tolerate priority inversions [48].

7.3.2 Base Rate Message

Next, we leverage the fact that only the TxTask is responsible for enqueuing the messages from the same node and, consequently, all messages with harmonic periods are enqueued exactly at the same time at each node. In most cases, there is one such message that runs at the *base rate* of the node, that is, at a period that is the greatest common divisor (*gcd*) of the periods of the messages, and has (typically) highest local priority. This message is enqueued every time any other message from the same node is added to the queue. This means that the arrival times of all the other messages must always match one of the arrival times of the base rate message. The base rate message functions as a reference of the arrival times of all the other messages. When such a message is not available, it is possible to build a virtual one, which is queued with the same period of the TxTask. If this correspondence among the arrival times can be reconstructed, then the arrival times of the other messages can be assigned by matching those of the corresponding base rate message.

7.3.3 Reconstruction of the True Message Periods and Arrival Times

The procedure for the reconstruction of the arrival times of messages is the following. First, the trace is analyzed for the detection of the actual period of each message, which may be different from the expected one, because of clock drifts.

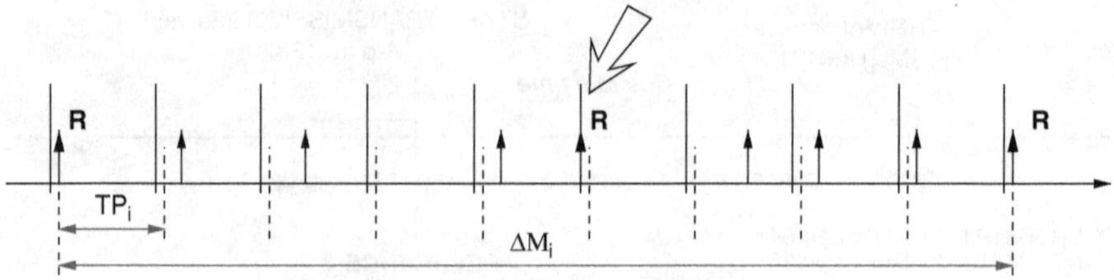

Fig. 7.4 Defining a reference periodic base for the arrival times

Then, *reference* message transmissions are used as a baseline for a procedure that assigns queuing times to all the message events in the trace. Before describing the details of the procedure, however, it is important to note that the procedure only needs to be applied to the base rate message. The arrival events of other messages will be obtained with a different method based on their phase detection.

The assignment of arrival times to the base rate message is performed as follows. The procedure first looks for the first and the last receive events of the reference message in the trace (see for example, Fig. 7.4 where the reference messages are marked with an R). Let us assume that all the trace events for the reference message m_i are labeled as $e_{i,0}, e_{i,2}, \ldots, e_{i,n}$. For the two reference events $e_{i,k}$ and $e_{i,j}$, with $j > k$, the corresponding time stamps are $t_{i,k}$ and $t_{i,j}$. Note that not all the trace events from m_i are reference message transmissions, thus it is not necessarily true that $k = 0$ or $j = n$. The time difference between these two reference transmissions is $\Delta t_i = t_{i,j} - t_{i,k}$. The number of message transmissions between the two reference events is $j - k$ and the actual (or true) message period is defined as

$$TP_i = \frac{\Delta t_i}{j - k}$$

For example, in the trace of the system in Table 7.3, the first reference event for message **0x0F1** is at time 3,500 (in μs), and the last at time 2,993,550 for the 300th message transmission event. The average true period for **0x0F1** is therefore $(2{,}993{,}550 - 3{,}500)/299 = 10{,}000.167$, which is a good match of the expected one at 10,000 μs. In the following step, starting from the time stamp of the first reference $t_{i,k}$, the tentative arrival times of all the other instances are defined as $a^*_{i,q} = t_{i,k} + TP_i \times p$ with $p = q - k$ (shown as dashed lines in Fig. 7.4).

Please note that the base rate message (in most cases) is also the highest priority message from each node and, therefore, the probability that it is transmitted after a bus idle time are much greater than other messages. However, it may be impossible, in some cases, to find two reference events in the trace. In this case, the first and last message transmission can simply be used in their place, with a reduced accuracy in the determination of the true period.

Table 7.1 True periods estimated for the messages of node E_{12}

id_i	Exp. $T_i\,(\mu s)$	True $T_i\,(\mu s)$	id_i	Exp. $T_i\,(\mu s)$	True $T_i\,(\mu s)$
0C1	10,000	10,618	1E5	10,000	10,617
0C5	10,000	10,618	1E9	20,000	21,234
17D	100,000	106,188	2F9	50,000	53,086
184	20,000	21,235			

7.3.4 *Queuing Jitter Because of TxTask Scheduling Delays*

The scheduling delay of TxTask can not be assumed to be negligible. The variable response time of the queuing task is inherited by the messages as their queueing jitter. At this point, the intermediate transmission events of the message need to be checked for the possible case in which the first event occurred with some amount of queuing jitter from the TxTask. For all the reference message events, the value $l^*_{i,q} = t_{i,q} - a^*_{i,q}$ is computed as an estimate of the *queuing jitter*, and the minimum value $\delta_{a^*_i} = \min_q\{l^*_{i,q}\}$ is applied as a correction to obtain the estimated arrival events $a_{i,q} = a^*_{i,q} + \delta_{a^*_i}$. In Fig. 7.4, a minimum value lower than zero is obtained for the event indicated by the arrow, and the arrival times are shifted left to the gray lines, which is the final result. The maximum value $t_{i,q} - a_{i,q}$ for all the reference messages from the same node is assumed as the maximum queuing jitter caused by the worst-case execution time of the TxTask.

7.3.5 *Clock Drifts and Actual Message Periods*

The procedure for computing the true period can be applied to all messages and can also be used to determine the clock drift at each node. For example, when applied to node E_{12} on the first case study (vehicle V_1, see Table 7.3), it gives the results of Table 7.1, which show that the hardware clock (or the signal activating the TxTask) has a drift of more than 6% for this node. The value is an actual clock drift, not a scheduling or response time jitter, as proved by the fact that it is almost constant across the entire message list from the same node. If the message trace is long, the estimates need to be corrected because the clock drift is in general not constant with time. Table 7.2 shows the effect of the clock drift on a message with id **0x128** from an experimental vehicle. The trace is recorded for more than 3,000 seconds. As in the table, there are about three more messages with frame ID **0x128** received in the first 1,000 s than in the second 1,000 s. Clearly, in this case the constant clock speed assumption, which translates in constant (base) message periods cannot be sustained. Otherwise, for example, the response time of message **0x128** (Fig. 7.5) is estimated as constantly growing, until it gets about three times its period.

Table 7.2 Evidence of clock drift on the same node: message **0x128** on an experimental vehicle with different number of frames received in the same length of time

Trace time (s)	# Received messages	True period (μs)
0–1,000	1,000,153	9,984.71
1,000–2,000	1,000,150	9,985.08

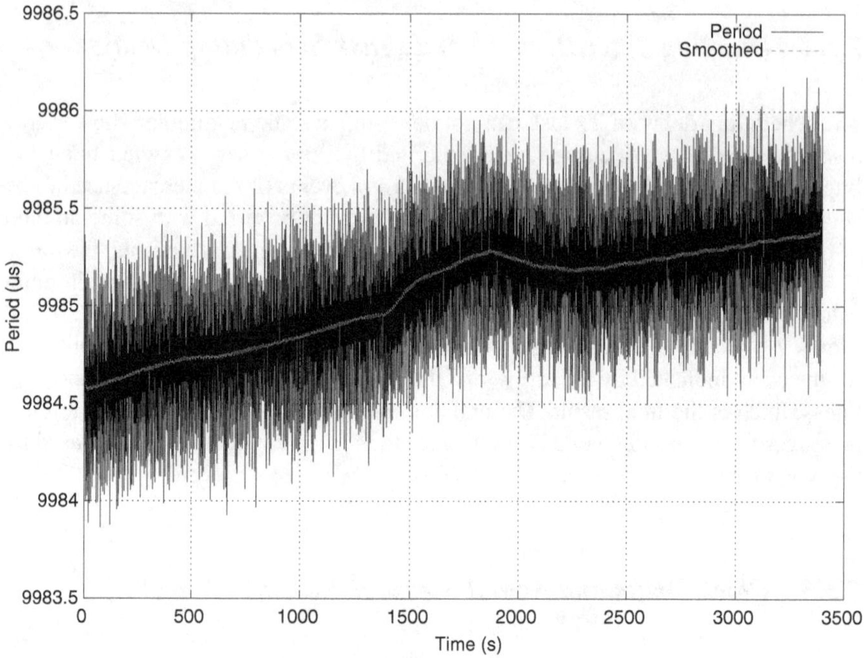

Fig. 7.5 Calculation of period for message **0x128** on an experimental vehicle considering clock drift

A simple heuristic to deal with this problem is to calculate the instant period at the j-th event of a base rate message m_i by considering the average period for k message transmission events around it

$$TP_{i,j} = \frac{t_{i,j+k/2} - t_{i,j-k/2}}{k} \tag{7.1}$$

The value of k can be assumed to be such that the clock speed variation on the same node is approximately linear. However, the time stamps $t_{i,j+k/2}$ and $t_{i,j-k/2}$ refer to the (end of) transmission events of the instances with index $j + k/2$ and $j - k/2$ of message m_i. Thus, the calculation of $TP_{i,j}$ in (7.1) is affected by the random variation of the response times. If $R_{i,j}$ is the average response time of the k messages sent for the events on which $TP_{i,j}$ is computed, then we have the following:

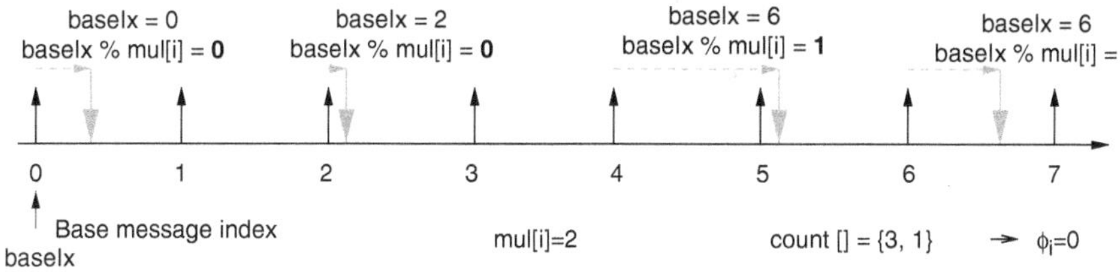

Fig. 7.6 Detection of the message phase

$$
\begin{aligned}
TP_{i,j} &= \frac{t_{i,j+k/2} - t_{i,j-k/2}}{k} \\
&= \frac{a_{i,j+k/2} + R_{i,j+k/2} - a_{i,j-k/2} - R_{i,j-k/2}}{k} \\
&= \frac{a_{i,j+k/2} - a_{i,j-k/2}}{k} + \frac{R_{i,j+k/2} - R_{i,j-k/2}}{k}
\end{aligned}
\tag{7.2}
$$

The first part on the right hand side $\frac{a_{i,j+k/2} - a_{i,j-k/2}}{k}$ can be considered as the true average period, but it is perturbed by the random variation of the noise $\mathcal{N} = \frac{R_{i,j+k/2} - R_{i,j-k/2}}{k}$. To demonstrate it, we plot the calculated $TP_{i,j}$ values for message **0x128** on an experimental vehicle as in Fig. 7.5, where $k = 1{,}000$, i.e. for a duration of approximately 10 s.

The calculation of the period is affected by the random variation in the term $\mathcal{N}$. However, this perturbed curve can be smoothed. The random term can be considered independent from $\frac{a_{i,j+k/2} - a_{i,j-k/2}}{k}$, with an average of zero. The 10,000-central moving average of $TP_{i,j}$ is shown in Fig. 7.5. This curve captures very well the trend of $TP_{i,j}$ caused by the variable clock drift.

7.3.6 Finding the Message Phase with Respect to the Base Rate Message

Assuming the existence of a message executing at the base rate for each node, the last step is to detect the phases of the other messages with respect to it. The procedure is formally described in Algorithm 4, and an example is provided in Fig. 7.6.

An index `baseIx` is associated with each transmission event of the base message, starting with 0. A multiplier expressing the period ratio for all the other messages with respect to the base message is stored in the array `mul[]`, with one entry for each message. The method is based on the initial assumption that the majority of the message instances is transmitted with a response time lower than the period of the base message. This assumption is valid in most cases, and exceptions can be detected a posteriori and corrected. More specifically, the

Algorithm 4 Calculate phase ϕ_i of $e_{i,j}$

```
for each trace event e = e_{i,j} do
    Message m = e.GetMessage();
    ECU c = m.GetEcu();
    Message bm = e.GetBaseRateMessage();
    if bm == m then
        c.SetBaseCount(index);
    else
        ph = c.GetBaseCount() % m.GetMultiplier();
        m.phase_count[ph]++;
    end if
end for
for each message m = m_i do
    m.phase = Index_of_max(m.phase_count);
end for
```

minimum response time of message m_i can be calculated as the minimum queuing delay which is due to the higher priority messages on the same node that are always queued before m_i, plus the transmission time of m_i itself. Then, the phase of the message should be adjusted such that the analyzed response time in the trace should be always larger than its calculated minimum response time. Finally, an array of counters `phase_count[]` of size `mul[i]` is associated to each message. For each transmission event of a message m_i with period larger than the base period, the algorithm finds the value of `baseIx` for the latest transmission of the base message, and computes the remainder of the division by the multiplier `val=baseIx % mul[i]`. Then, the counter with index matching the computed value is incremented. At the end, the position of the counter array with the largest value indicates the phase ϕ_i of the message.

In the figure, four transmission events have phase count of 0, and only one transmission event has a phase count of 1. Therefore, the algorithm assumes $\phi_i = 0$ and assigns the corresponding arrival times by back-traversing the event list until it finds the first queuing event of the base rate message with index $k \times mul + \phi_i$, with $k \in 0, 1, 2, \ldots$. If no such event exists, i.e., the base rate transmission of index ϕ_i occurs after the message transmission event that the algorithm is considering, then the arrival time of the base rate event with index ϕ_i is decremented by the period of the message, and the procedure is tried one more time (see Algorithm 5). The result of the phase assignment algorithm for our case study is shown in Tables 7.3 and 7.4, where T_i and ϕ_i are in ms.

7.4 Analysis of Automotive Case Studies

The trace analysis procedures described in this chapter have been applied to experimental examples: the first example contains a subset of vehicle V_1 with five nodes (Table 7.3) and a subset of vehicle V_2 with four nodes (Table 7.4),

Algorithm 5 Calculate arrival time of $e_{i,j}$

```
for each trace event e = e_{i,j} do
    Message m = e.GetMessage();
    ECU c = m.GetEcu();
    Message bm = e.GetBaseRateMessage();
    if bm == m then
        c.SetBaseCount(index);
        c.SetLastBaseArrival(e.GetArrivalTime());
    else
        mcnt = m.GetBaseDynCnt();
        bcnt = c.GetBaseCount();
        m.SetBaseDynCnt(mcnt + m.GetMultiplier());
        Δ = bcnt-mcnt;
        if Δ < 0 then
            Δ = Δ + m.GetMultiplier();
        end if
        e.SetArrivalTime(c.GetLastBaseArrival() - Δ×bm.GetTPeriod();
    end if
end for
```

Table 7.3 Nodes and messages from a subset of vehicle V_1

Υ_i^{src}	id_i	T_i	ϕ_i	Υ_i^{src}	id_i	T_i	ϕ_i	Υ_i^{src}	id_i	T_i	ϕ_i
E_{11}	0F1	10	0	E_{12}	0C1	10	0	E_{13}	4C1	500	12.5
E_{11}	120	5,000	1,490	E_{12}	0C5	10	0	E_{13}	4D1	500	12.5
E_{11}	12A	100	90	E_{12}	17D	100	70	E_{13}	4F1	1,000	12.5
E_{11}	130	1,000	490	E_{12}	184	20	0	E_{13}	772	1,000	12.5
E_{11}	138	1,000	490	E_{12}	1E5	10	0	E_{14}	324	500	30
E_{11}	1E1	30	20	E_{12}	1E9	20	0	E_{14}	524	10	0
E_{11}	1F1	100	90	E_{12}	2F9	50	0	E_{14}	528	10	0
E_{11}	1F3	20	10	E_{13}	0C9	12.5	0	E_{14}	77E	1,000	540
E_{11}	3C9	100	90	E_{13}	191	12.5	0	E_{15}	0F9	12.5	0
E_{11}	3F1	250	240	E_{13}	1A1	25	12.5	E_{15}	199	12.5	0
E_{11}	4E1	1,000	510	E_{13}	3C1	100	12.5	E_{15}	19D	25	0
E_{11}	4E9	1,000	510	E_{13}	3D1	100	12.5	E_{15}	1F5	25	0
E_{11}	514	1,000	520	E_{13}	3E9	100	12.5	E_{15}	4C9	500	0
E_{11}	52A	1,000	520	E_{13}	3F9	250	12.5	E_{15}	77F	1,000	0
E_{11}	771	1,000	520								

where trace analysis is used to detect the source of a timing anomaly causing a message overwrite on the receiving side, as shortly outlined in the previous sections. The second example contains a bus from a hybrid vehicle, where trace analysis helps in the reconstruction of the message-to-ECU relation (the information was not available). More deviations from the assumptions made in the worst-case analysis (Chap. 3 and [60]) and the stochastic analysis (Chap. 4 and [65]) are found, including the existence of priority inversion of messages on the same node, and the existence of a copy time of messages from the queue to the transmit register(s) at the source node larger than the transmission time of the interframe bits (which introduces the probability of additional blocking from lower priority messages from other nodes).

Table 7.4 Nodes and messages from a subset of vehicle V_2

Υ_i^{src}	id_i	T_i	ϕ_i	Υ_i^{src}	id_i	T_i	ϕ_i	Υ_i^{src}	id_i	T_i	ϕ_i
E_{21}	140	20	0	E_{22}	380	20	10	E_{23}	300	100	0
E_{21}	144	20	10	E_{22}	388	20	10	E_{23}	308	100	70
E_{21}	180	10	0	E_{22}	38A	1,500	1,250	E_{23}	348	250	10
E_{21}	280	30	20	E_{22}	390	1,000	250	E_{23}	410	100	90
E_{21}	2F0	100	0	E_{22}	670	1,200	50	E_{23}	510	500	450
E_{21}	330	100	20	E_{22}	674	1,200	50	E_{24}	150	12.5	0
E_{21}	420	100	40	E_{23}	110	10	0	E_{24}	151	12.5	0
E_{22}	128	10	0	E_{23}	120	10	0	E_{24}	320	100	62.5
E_{22}	130	200	0	E_{23}	124	10	0	E_{24}	520	100	0
E_{22}	2D0	500	250	E_{23}	170	500	350				

7.4.1 Overwrite Error at the Destination Node for Message 0x1F3 in V_1

Figures 7.7 and 7.8 show the results of the analysis for two sets of messages. Two nodes, E_{11} and E_{12}, are considered for reasons that will be clear in the following. In all graphs, two curves are plotted. One curve shows the density (in light blue) indicating the number of message instances that were delivered with a response time equal to the value on the x-axis. The other curve (in black) indicates the cumulative fraction of messages delivered with latency lower than or equal to x. The response time of the highest priority messages on both nodes are similar. The overall response time is generally small, with values smaller than 850 μs. In both cases, the curve of the cumulative fraction of messages with a response time lower than or equal to x is almost continuous, with a steep rise and rapidly approaching the unity value.

When plotting the response time values of medium and low priority messages from node E_{12}, a similar shape of the curve is found, as shown in Fig. 7.7. The cumulative fraction still rises continuously. The cumulative function starts rising from larger response time values (around 500 μs for **0x184**), which gives an indication of the minimum queuing delay caused by the local higher priority messages that are *always* enqueued together with the considered message. The slope of the curve is different, steeper for the higher priority message, which suffers a smaller arbitration delay from remote messages and more flat for the lower priority message. In general, this is the behavior we would expect for messages sent by nodes in which the TxObject is managed by interrupt (see also [65]).

A completely different shape is obtained for the response time graphs of the messages output by node E_{11}, as shown in Fig. 7.8. The cumulative graph is a staircase, and the density graph is highly discontinuous, showing clusters at response time values that are approximately multiples of a given value, in this case 2,500 μs. This is a clear indication of a node, in which the assignment of the TxObject is managed by a polling task. We know from previous discussions with the supplier of the node that the period of the polling task was in this case actually equal to 2,500 μs, therefore confirming the trace analysis results.

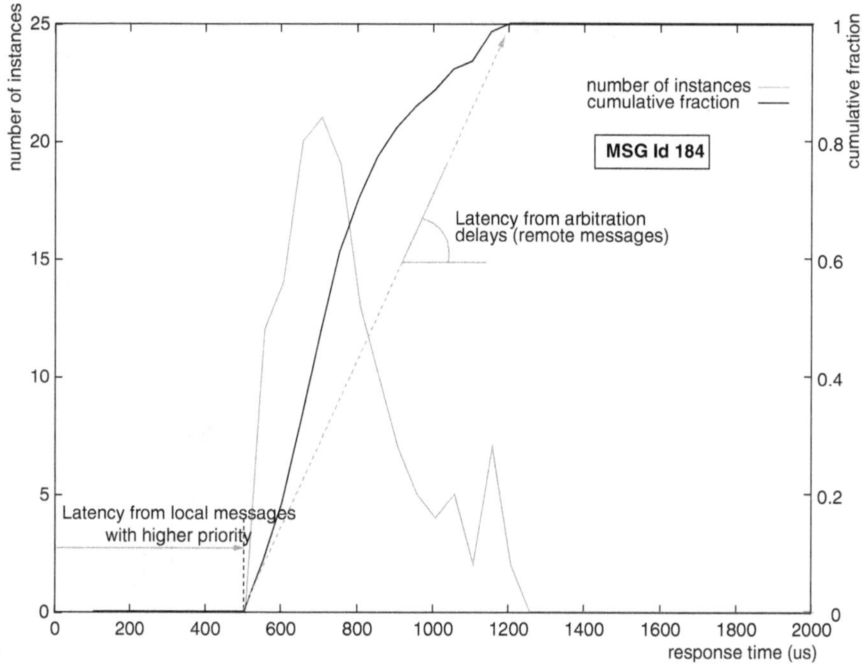

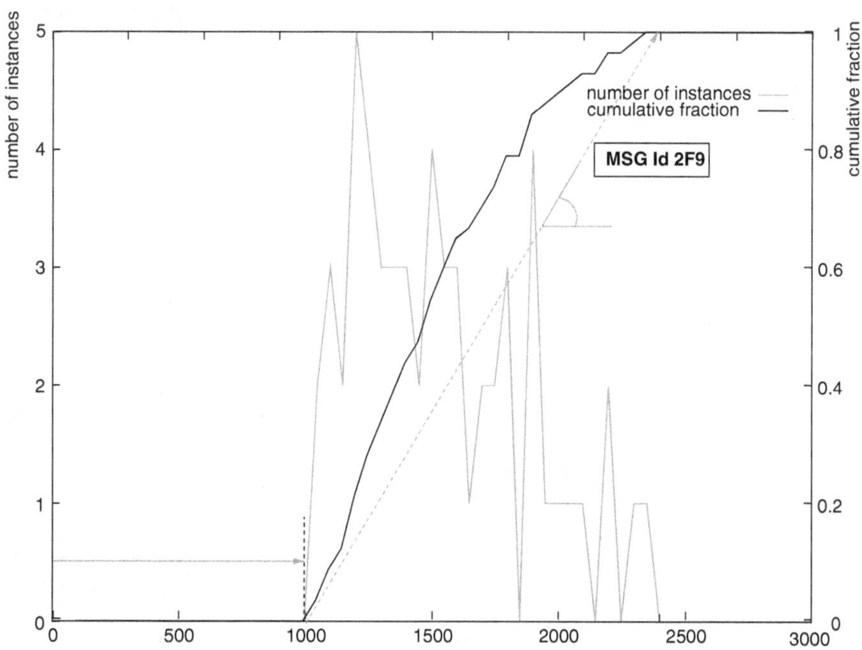

Fig. 7.7 Response time of messages **0x184** and **0x2F9** on E_{12} of vehicle V_1

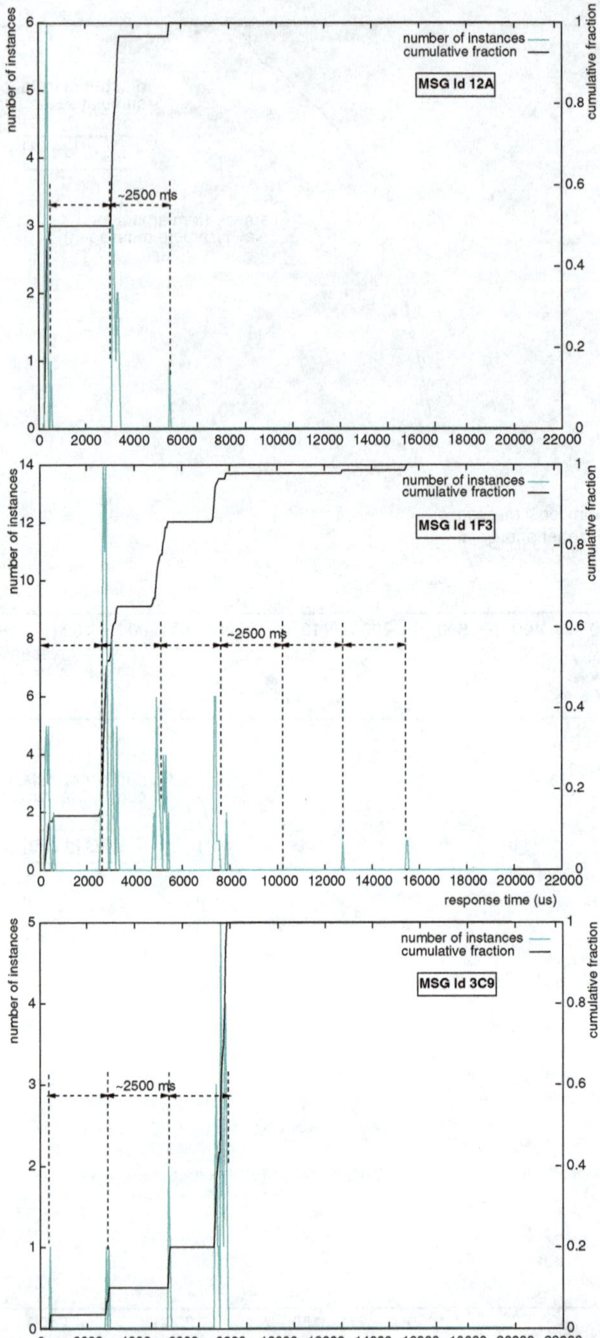

Fig. 7.8 Response time of messages **0x12A**, **0x1F3** and **0x3C9** on E_{11} of vehicle V_1

Table 7.5 Trace segment with arrival times and response times

1 ▷	t=1,493,860	Id:0F1	Len:32	Arr:1,493,440	Resp:420	
	t=1,494,120	Id:0C5	Len:64	Arr:1,493,620	Resp:500	
	t=1,494,350	Id:184	Len:48	Arr:1,493,620	Resp:730	
	t=1,494,640	Id:1E5	Len:64	Arr:1,493,620	Resp:1,020	
	t=1,494,850	Id:0C9	Len:56	Arr:1,494,352	Resp:498	
	t=1,495,090	Id:1E9	Len:64	Arr:1,493,620	Resp:1,470	
	t=1,495,320	Id:191	Len:64	Arr:1,494,352	Resp:968	
	t=1,495,510	Id:2F9	Len:40	Arr:1,493,620	Resp:1,890	
	t=1,495,650	Id:1A1	Len:24	Arr:1,494,352	Resp:1,298	
	t=1,495,860	Id:1C3	Len:40	Arr:1,495,109	Resp:751	
2 ▷	t=1,496,810	Id:120	Len:40	Arr:1,493,440	Resp:3,370	REF
3 ▷	t=1,499,010	Id:12A	Len:64	Arr:1,493,440	Resp:5,570	REF
4 ▷	t=1,500,890	Id:130	Len:40	Arr:1,493,440	Resp:7,450	REF
5 ▷	t=1,501,240	Id:138	Len:40	Arr:1,493,440	Resp:7,800	REF
	t=1,501,860	Id:0F9	Len:64	Arr:1,501,718	Resp:142	REF
	t=1,502,060	Id:199	Len:64	Arr:1,501,718	Resp:342	
	t=1,502,410	Id:524	Len:64	Arr:1,502,280	Resp:130	REF
	t=1,502,590	Id:528	Len:40	Arr:1,502,280	Resp:310	
	t=1,503,560	Id:0F1	Len:32	Arr:1,503,440	Resp:120	REF
6 ▷	t=1,503,750	Id:1E1	Len:24	Arr:1,493,440	Resp:10,310	
	t=1,504,410	Id:0C1	Len:64	Arr:1,504,238	Resp:172	REF
	t=1,504,650	Id:0C5	Len:64	Arr:1,504,238	Resp:412	
	t=1,504,900	Id:1C7	Len:56	Arr:1,504,830	Resp:70	
	t=1,505,170	Id:1E5	Len:64	Arr:1,504,238	Resp:932	
7 ▷	t=1,506,360	Id:1F1	Len:64	Arr:1,493,440	Resp:12,920	REF
	t=1,507,960	Id:0C9	Len:56	Arr:1,506,850	Resp:1,110	REF
	t=1,508,190	Id:191	Len:64	Arr:1,506,850	Resp:1,340	
8 ▷	*t=1,508,910*	*Id:1F3*	*Len:16*	*Arr:1,493,440*	*Resp:15,470*	REF
9 ▷	t=1,511,250	Id:3C9	Len:64	Arr:1,493,440	Resp:17,810	REF
	t=1,512,390	Id:524	Len:64	Arr:1,512,280	Resp:110	REF
	t=1,512,570	Id:528	Len:40	Arr:1,512,280	Resp:290	
1 •	t=1,513,580	Id:0F1	Len:32	Arr:1,513,440	Resp:140	REF
2 •	*t=1,513,780*	*Id:1F3*	*Len:16*	*Arr:1,513,440*	*Resp:340*	

Finally, after the reconstruction of the arrival times and the corresponding response times, Table 7.5 shows the trace annotated with the additional information (arrival, activation and response times are in μs), so that the sequence of events that led to the large response time of message **0x1F3** can be understood and the error possibly recovered. The sequence that leads to the timing error is actually caused by the simultaneous queuing of all the messages with priority higher than **0x1F3** at time 1,493,440. These messages are (in priority order) **0x0F1, 0x120, 0x12A, 0x130, 0x138, 0x1E1** and **0x1F1**, and are copied in the TxObject by a polling task that forces a time separation of 2.5 ms between their transmissions. This is confirmed by the existence of bus idle time prior to the transmission of most of them (the transmission events of these messages are indicated by horizontal lines in the figure). The instance of **0x1F3** that is enqueued at time 1,493,440 is transmitted at 1,508,910, with an estimated response time of 15,470 μs. The following instance, however, is enqueued right after the base rate message at 1,513,580 and transmitted immediately, at time 1,513,780, with a response time of only 340 μs. The absence of a response time of at least one period of the polling task can be explained considering that **0x0F1** may have been written in the TxObject by the TxTask. When the polling task is activated immediately after **0x0F1** has already been transmitted, the TxObject buffer is available for **0x1F3**.

Next, we will discuss how to find errors in the result of the phase detection algorithm and how to fix them. With reference to Table 7.3, consider the messages **0x3F1** and **0x4E1** from the node E_{11} on vehicle V_1. The period of **0x4E1** is a divisor of the period of **0x3F1**, and the two messages must always be enqueued together. This implies that the response time of **0x4E1** is always larger than **0x3F1**. Given that the response time of **0x3F1** is as large as 24,890, we expect that the response time values between 5,000 and 6,000 that are obtained for **0x4E1**, should actually be corrected by adding twice the period of the base message (+20,000). This would also put the phase of **0x4E1** at 49, which is same as the phase of the other higher priority messages with the same rate from the same node, confirming the intuition on the choice of the designers to send all the messages from E_{11} with the same phase. A similar reasoning applies to the other lower priority messages from E_{11}. It should be noted that the choice of sending all messages with the same phase is one major cause (but not the only one) for the occurrence of events like the error shown in Table 7.5 that leads to the large response time of message **0x1F3**.

7.4.2 Aperiodic Message Transmission

In our analysis of the trace, we assume all the messages are periodic, meaning that their transmission is requested on a periodic base with a possible queuing jitter. Accordingly, we would expect these messages to arrive at the destination nodes with the same periodic base and an even larger jitter because of the variability of their response times. However, this is not always true in our experiments. The aperiodic messages are **0x130** from node E_{22} and **0x300** from node E_{23}, both in vehicle V_2 (Table 7.4). Figure 7.9 shows a trace segment with the receive time stamps for message **0x130**. There are two sets of activations for **0x130**, one with a period of about *200* ms, the other with period of about *1,500* ms, as indicated by the triangles. In Fig. 7.10 the trace snapshot of message **0x300** is depicted, with two sources of periodic transmission events, having the same period of *100* ms and a relative phase equal to *20* ms.

7.4.3 Local Priority Inversion

The assumption of a priority-based management at each node for the outgoing messages does not always hold in actual implementations. This leads to aberrations

```
  t= 162263 Id:130 Len:48 Data:  00 00 00 71 AD 2B
  t= 361954 Id:130 Len:48 Data:  00 00 00 71 AD 2B
▷ t= 392140 Id:130 Len:48 Data:  00 00 00 71 AD 2B
  t= 561583 Id:130 Len:48 Data:  00 00 00 71 AD 2B
  t= 761282 Id:130 Len:48 Data:  00 00 00 71 AD 2B
  t= 960986 Id:130 Len:48 Data:  00 00 00 71 AD 2B
  t=1160662 Id:130 Len:48 Data:  00 00 00 71 AD 2B
  t=1360794 Id:130 Len:48 Data:  00 00 00 71 AD 2B
  t=1560506 Id:130 Len:48 Data:  00 00 00 71 AD 2B
  t=1759929 Id:130 Len:48 Data:  00 00 00 71 AD 2B
▷ t=1890107 Id:130 Len:48 Data:  00 00 00 71 AD 2B
  t=1959622 Id:130 Len:48 Data:  00 00 00 71 AD 2B
  t=2159314 Id:130 Len:48 Data:  00 00 00 71 AD 2B
  t=2358862 Id:130 Len:48 Data:  00 00 00 71 AD 2B
  t=2558671 Id:130 Len:48 Data:  00 00 00 71 AD 2B
  t=2758647 Id:130 Len:48 Data:  00 00 00 71 AD 2B
  t=2958071 Id:130 Len:48 Data:  00 00 00 71 AD 2B
  t=3157783 Id:130 Len:48 Data:  00 00 00 71 AD 2B
  t=3357789 Id:130 Len:48 Data:  00 00 00 71 AD 2B
▷ t=3387786 Id:130 Len:48 Data:  00 00 00 71 AD 2B
  t=3557068 Id:130 Len:48 Data:  00 00 00 71 AD 2B
```

Fig. 7.9 An aperiodic message **0x130** from E_{22} on vehicle V_2: activation interval in general is 200 ms, with additional activations every 1,500 ms indicated by the triangles

```
▷ t= 48518 Id:300 Len:64 Data:  04 81 00 61 43 00 00 00
⇒ t= 68457 Id:300 Len:64 Data:  04 81 00 61 43 00 00 00
▷ t=148464 Id:300 Len:64 Data:  04 81 00 61 45 00 00 00
⇒ t=168481 Id:300 Len:64 Data:  04 81 00 61 45 00 00 00
▷ t=248468 Id:300 Len:64 Data:  04 81 00 61 46 00 00 00
⇒ t=268457 Id:300 Len:64 Data:  04 81 00 61 46 00 00 00
▷ t=348778 Id:300 Len:64 Data:  04 81 00 61 49 00 00 00
⇒ t=368462 Id:300 Len:64 Data:  04 81 00 61 49 00 00 00
▷ t=448471 Id:300 Len:64 Data:  04 81 00 61 4A 00 00 00
⇒ t=468466 Id:300 Len:64 Data:  04 81 00 61 4A 00 00 00
▷ t=548477 Id:300 Len:64 Data:  04 81 00 61 4C 00 00 00
⇒ t=569464 Id:300 Len:64 Data:  04 81 00 61 4C 00 00 00
▷ t=648524 Id:300 Len:64 Data:  04 81 00 61 4E 00 00 00
⇒ t=668465 Id:300 Len:64 Data:  04 81 00 61 4E 00 00 00
▷ t=748481 Id:300 Len:64 Data:  04 81 00 61 50 00 00 00
⇒ t=768471 Id:300 Len:64 Data:  04 81 00 61 50 00 00 00
▷ t=849041 Id:300 Len:64 Data:  04 81 00 61 51 00 00 00
⇒ t=868709 Id:300 Len:64 Data:  04 81 00 61 51 00 00 00
```

Fig. 7.10 An aperiodic message **0x300** from E_{23} on vehicle V_2: two sets of periodic transmissions at nominal period 100 ms indicated by *triangles* and *arrows* respectively, with relative phase = 80 ms/20 ms

```
  t=1857316 Id:110 Len:64 Data:  00 09 BF 00 00 06 00 00
  t=1857548 Id:120 Len:64 Data:  03 85 23 83 06 EA 03 85
▷ t=1857696 Id:170 Len:24 Data:  01 00 86
  t=1858256 Id:124 Len:40 Data:  00 03 83 03 85
  ......
  t=3877361 Id:110 Len:64 Data:  00 09 C4 00 00 06 00 00
  t=3877597 Id:120 Len:64 Data:  03 83 23 81 06 EA 03 82
▷ t=3877819 Id:308 Len:56 Data:  00 80 2A 00 00 00 AD
  t=3878309 Id:124 Len:40 Data:  00 03 81 03 83
  ......
  t=4017366 Id:110 Len:64 Data:  00 09 C4 00 00 06 00 00
  t=4017600 Id:120 Len:64 Data:  03 85 23 80 06 EA 03 81
▷ t=4017768 Id:348 Len:32 Data:  08 48 43 FF
  t=4018312 Id:124 Len:40 Data:  00 03 80 03 85
```

Fig. 7.11 Local priority inversion of message **0x124** on node E_{23} of vehicle V_2, as shown in the trace segment. Note that on node E_{23} the period of TxTask should be *10* ms, the *gcd* of the periods. **0x124** should always be queued together with **0x170, 0x308, 0x348, 0x410** and **0x510**, but it is transmitted after **0x170, 0x308, 0x348** as indicated by the *triangles*

```
  t=3211141 Id:0F9 Len:64 Data:  00 00 40 00 00 00 03 FF
▷ t=3211282 Id:4C9 Len:32 Data:  40 6D 00 00
  t=3211516 Id:199 Len:64 Data:  0F FF 0E 70 F1 90 00 FF
  ......
  t=9711519 Id:0F9 Len:64 Data:  00 00 40 00 00 00 03 FF
▷ t=9711687 Id:4C9 Len:32 Data:  40 6D 00 00
  t=9711921 Id:199 Len:64 Data:  0F FF 0E 70 F1 90 00 FF
  ......
  t=10211571 Id:0F9 Len:64 Data:  00 00 40 00 00 00 03 FF
▷ t=10211739 Id:4C9 Len:32 Data:  40 6D 00 00
  t=10211973 Id:199 Len:64 Data:  0F FF 0E 70 F1 90 00 FF
```

Fig. 7.12 Local priority inversion of message **0x199** on node E_{15} of vehicle V_1, as shown in the trace segment. Note that on node E_{15} the period of TxTask should be *12.5* ms, the *gcd* of the periods. **0x199** should always be queued together with **0x4C9**, but it is transmitted after **0x4C9** as indicated by the *triangles*

such that a message may be transmitted after some lower priority messages on the same node even if it is queued along with these lower priority message. Figure 7.11 shows one such example message **0x124** from node E_{23} on vehicle V_2. As on the top of Fig. 7.11, the period of the TxTask on E_{23} should be the *gcd* of the message periods, i.e. *10* ms. Message **0x124** with the same period as TxTask is queued every time messages **0x170, 0x308, 0x348, 0x410** and **0x510** are queued. Thus, we would expect it to be transmitted before these lower priority messages if the driver queuing policy on node E_{23} is priority-based. However, as shown in the trace segment of Fig. 7.11, **0x124** sometimes is transmitted after **0x170, 0x308, 0x348** as indicated by the triangles in the figure. Similarly, on node E_{15} of vehicle V_1, messages **0x199, 0x19D** and **0x1F5** experience priority inversion by local lower priority messages, as shown in Figs. 7.12–7.14 respectively.

One possible cause to this local priority inversion may be the existence of two separate queues of messages and two TxObjects used for transmission (one for each

```
   t= 211116 Id:0F1 Len:32 Data:  34 02 00 40
   t= 211350 Id:199 Len:64 Data:  0F FF 0E 70 F1 90 00 FF
▷  t= 211518 Id:4C9 Len:32 Data:  40 6D 00 00
   t= 211764 Id:19D Len:64 Data:  00 00 00 00 00 09 A1 FF
   t= 211988 Id:1F5 Len:56 Data:  0F 0F 00 01 00 00 00
   ......
   t= 711252 Id:0F1 Len:32 Data:  1C 02 00 40
   t= 711486 Id:199 Len:64 Data:  0F FF 0E 70 F1 90 00 FF
▷  t= 711655 Id:4C9 Len:32 Data:  40 6D 00 00
   t= 711899 Id:19D Len:64 Data:  00 00 00 00 00 0A 3E FF
   ......
   t=2711191 Id:0F9 Len:64 Data:  00 00 40 00 00 00 03 FF
▷  t=2711359 Id:4C9 Len:32 Data:  40 6D 00 00
   t=2711529 Id:0F1 Len:32 Data:  1C 02 00 40
   t=2711764 Id:199 Len:64 Data:  0F FF 0E 70 F1 90 00 FF
▷  t=2711994 Id:77F Len:56 Data:  00 00 00 00 00 00 00
   t=2712236 Id:19D Len:64 Data:  00 00 00 00 00 0A 21 FF
```

Fig. 7.13 Local priority inversion of message **0x19D** on node E_{15} of vehicle V_1, as shown in the trace segment. **0x19D** should always be queued together with **0x4C9** and **0x77F**, but it is transmitted after **0x4C9** or **0x77F** as indicated by the *triangles*

```
   t= 711486 Id:199 Len:64 Data:  0F FF 0E 70 F1 90 00 FF
▷  t= 711655 Id:4C9 Len:32 Data:  40 6D 00 00
   t= 711899 Id:19D Len:64 Data:  00 00 00 00 00 0A 3E FF
▷  t= 712129 Id:77F Len:56 Data:  00 00 00 00 00 00 00
   t= 712351 Id:1F5 Len:56 Data:  0F 0F 00 01 00 00 00
   ......
   t=3711718 Id:0F1 Len:32 Data:  1C 02 00 40
   t=3711964 Id:19D Len:64 Data:  00 00 00 00 00 0A 0E FF
▷  t=3712194 Id:77F Len:56 Data:  00 00 00 00 00 00 00
   t=3712416 Id:1F5 Len:56 Data:  0F 0F 00 01 00 00 00
```

Fig. 7.14 Local priority inversion of message **0x1F5** on node E_{15} of vehicle V_1, as shown in the trace segment. **0x1F5** should always be queued together with **0x4C9** and **0x77F**, but it is transmitted after **0x4C9** or **0x77F** as indicated by the *triangles*

queue), and a finite copy time between the message queue and the TxObjects. On node E_{23} of vehicle V_2, messages **0x110**, **0x120** and **0x124** belong to one driver-level queue, while messages **0x170**, **0x308** and **0x348** are managed using another queue, and one TxObject is dedicated to serve the messages out of each queue. As an example, Fig. 7.15 shows the sequence of events leading to the (limited) priority inversion of message **0x124**. In Step (a), messages **0x120**, **0x124**, **0x170**,**0x308** and **0x348** are activated, and both TxObjects are empty, thus **0x120** and **0x170** are copied into them. Some time later, message **0x120** wins the arbitration and gets transmitted on the bus (Step (b)). After **0x120** finishes its transmission, an interrupt signal is thrown and the interrupt handler will be called to copy the next message available, **0x124**, into the TxObject. If the copy time between the driver queue and the TxObject is larger than the time used by the 11 inter-frame bits on the bus, then the next arbitration round starts before **0x124** is copied into the TxObject. If **0x170** is the highest priority message in the system, it will win the access to the bus and bring an additional waiting time to **0x124**.

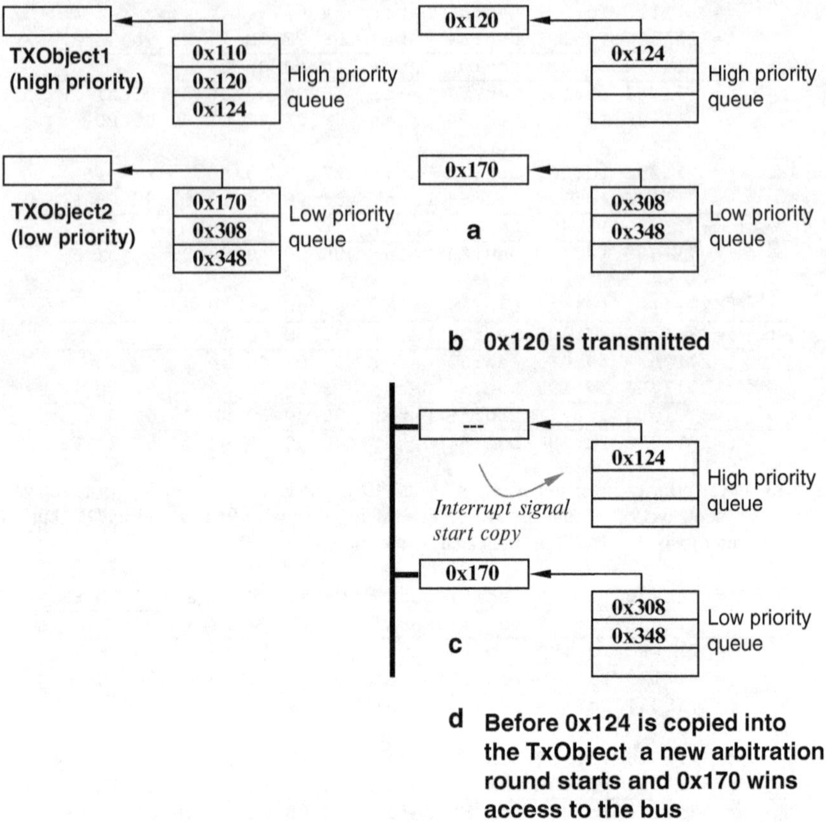

Fig. 7.15 Interpretation of possible cause of local priority inversion for message **0x124**

Similar arguments can be made for node E_{15} of vehicle V_1, where messages **0x0F9**, **0x199**, **0x19D** and **0x1F5** belong to one queue, and messages **0x4C9** and **0x77F** belong to a different queue. As a result, messages **0x199**, **0x19D** and **0x1F5** may suffer local priority inversion from **0x4C9** and **0x77F**, as in Figs. 7.12–7.14.

7.4.4 Remote Priority Inversion

Similar to the previous case, when the copy time is larger than the 11 bits inter-frame time, a remote message with lower priority gets a chance to be transmitted. Figure 7.16 shows an example of such remote priority inversion. On node E_{24} of vehicle V_2, messages **0x150** and **0x151** have the same periods as the TxTask, and are always queued together. In some of the trace segment, messages **0x380**, **0x388**, and **0x410** from other nodes in the system sneak in, and get transmitted between

```
   t=222236 Id:150 Len:64 Data:  40 00 09 60 3F FF F6 9F
▷  t=222527 Id:380 Len:64 Data:  09 42 20 00 70 40 FC BF
   t=222766 Id:151 Len:64 Data:  00 FF 09 22 00 00 0F 3F
   ......
   t=297743 Id:150 Len:64 Data:  C0 00 09 60 3F FD F6 9D
▷  t=297989 Id:410 Len:64 Data:  00 00 00 96 2B 00 00 00
   t=298229 Id:151 Len:64 Data:  00 FF 09 25 00 00 0F 3F
   ......
   t=322497 Id:150 Len:64 Data:  40 00 09 60 3F FF F6 9F
▷  t=322733 Id:388 Len:64 Data:  21 12 68 19 00 00 DC 80
   t=322978 Id:151 Len:64 Data:  00 FF 09 21 00 00 0F 3F
```

Fig. 7.16 Remote priority inversion of message **0x151** on E_{24} of vehicle V_2. **0x151** should always be queued together with **0x150**, but **0x380**, **0x388**, and **0x410** get transmitted between **0x150** and **0x151**, as indicated by the *triangles*

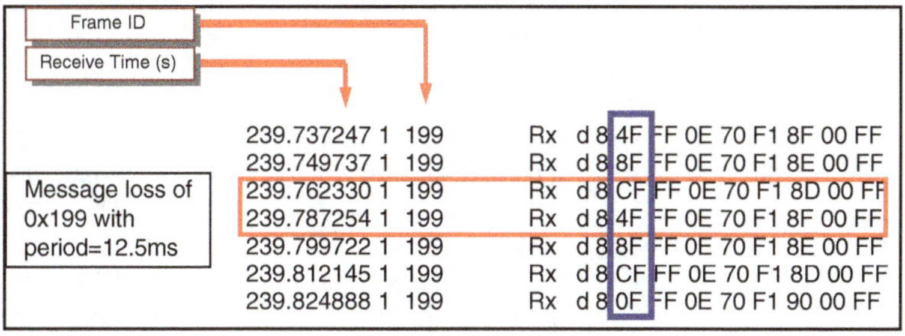

Fig. 7.17 Trace segment of message loss at the transmission node: **0x199** from E_{15} on vehicle V_1

0x150 and **0x151**. This is evidence that the copy time from the driver queue to the TxObject is larger than the inter-frame time: after **0x150** is transmitted, a new arbitration round starts before **0x151** is copied into the TxObject, such that remote lower priority messages, e.g., messages **0x380**, **0x388**, and **0x410**, get a chance to access the bus and lead to priority inversion for **0x151**.

Of course, theoretically message **0x151** may even experience an additional delay because of remote priority inversion. For example, in the case of Fig. 7.16 a set of messages with priority higher than **0x151** may get into the TxObjects during the transmission of **0x380** (or **0x388**), thus interfering with the transmission of message **0x151** for a longer time. Although intuitively the probability of this case is very low and will not have a big impact on the distribution of the message response time, it certainly introduces an additional difficulty to formally compute the worst case message response time.

7.4.5 Message Loss at Source Node

Another error that occurs in the system is the omission of message transmissions at the source side. Figure 7.17 shows the trace segment for message **0x199** from node E_{15} on vehicle V_1, for which the period is configured at *12.5* ms. An instance

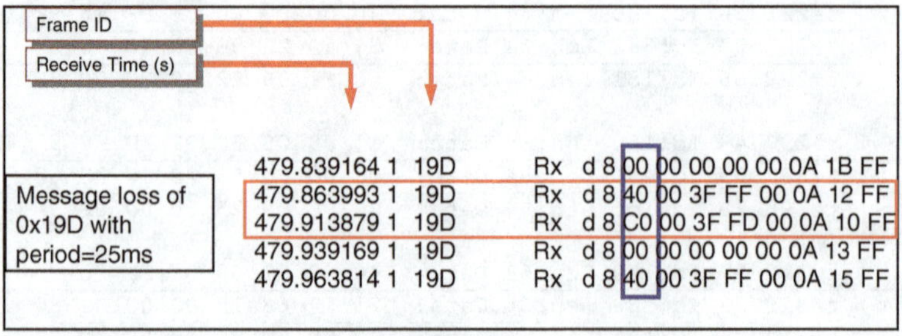

Fig. 7.18 Trace segment of message loss at the transmission node: **0x19D** from E_{15} on vehicle V_1

Table 7.6 Messages from the hybrid vehicle (periods in *ms*)

m_i	Υ_i^{src}	T_i	True T_i	Deviation	m_i	Υ_i^{src}	T_i	True T_i	Deviation
m_1	E_{06}	12.5	12.50434	1.000347	m_{28}	E_{00}	100	100.1435	1.001435
m_2	E_{06}	12.5	12.50465	1.000372	m_{29}	E_{00}	1,000	1,001.436	1.001436
m_3	E_{06}	12.5	12.50434	1.000347	m_{30}	E_{00}	1,000	1,001.436	1.001436
m_4	E_{06}	12.5	12.50465	1.000372	m_{31}	E_{00}	1,000	1,001.435	1.001435
m_5	E_{06}	50	50.01078	1.000216	m_{32}	E_{03}	20	19.9928	0.99964
m_6	E_{06}	50	50.01077	1.000215	m_{33}	E_{03}	50	49.98202	0.99964
m_7	E_{00}	50	50.11753	1.002351	m_{34}	E_{03}	500	499.8201	0.99964
m_8	E_{00}	100	100.1435	1.001435	m_{35}	E_{03}	20	19.9928	0.99964
m_9	E_{00}	20	20.02871	1.001435	m_{36}	E_{03}	500	499.8198	0.99964
m_{10}	E_{00}	50	50.07175	1.001435	m_{37}	E_{03}	20	19.9928	0.99964
m_{11}	E_{00}	125	125.1794	1.001435	m_{38}	E_{00}	200	200.2872	1.001436
m_{12}	E_{06}	25	25.00866	1.000346	m_{39}	E_{03}	1,000	999.6396	0.99964
m_{13}	E_{06}	10,000	10,004.151	1.000415	m_{40}	E_{04}	100	100.235	1.0023
m_{14}	E_{00}	100	100.1435	1.001435	m_{41}	E_{04}	100	100.235	1.00235
m_{15}	E_{02}	100	99.83493	0.998349	m_{42}	E_{04}	100	100.235	1.00235
m_{16}	E_{00}	100	100.1435	1.001435	m_{43}	E_{04}	50	50.11756	1.002351
m_{17}	E_{02}	100	99.83493	0.998349	m_{44}	E_{04}	50	50.11752	1.00235
m_{18}	E_{00}	1,000	1,001.436	1.001436	m_{45}	E_{04}	50	50.11758	1.002352
m_{19}	E_{00}	1,000	1,001.436	1.001436	m_{46}	E_{04}	50	50.11752	1.00235
m_{20}	E_{00}	1,000	1,001.436	1.001436	m_{47}	E_{05}	1,000	1,024.314	1.024314
m_{21}	E_{01}	1,000	1,000.524	1.000524	m_{48}	E_{05}	1,000	1,024.314	1.024314
m_{22}	E_{01}	500	500.2619	1.000524	m_{49}	E_{05}	1,000	1,024.314	1.024314
m_{23}	E_{01}	500	500.262	1.000524	m_{50}	E_{05}	1,000	1,024.314	1.024314
m_{24}	E_{01}	500	500.262	1.000524	m_{51}	E_{05}	125	128.039	1.024312
m_{25}	E_{01}	500	500.262	1.000524	m_{52}	E_{05}	125	128.039	1.024312
m_{26}	E_{00}	100	100.1436	1.001436	m_{53}	E_{05}	125	128.0389	1.024312
m_{27}	E_{00}	100	100.1436	1.001436	m_{54}	E_{00}	1,000	1,001.436	1.001436

of **0x199** is expected to be transmitted on the bus at around 239.775 s. Similarly, in Fig. 7.18 an instance of **0x199** is expected around 479.889 s. However, in the respective trace segments, the tracing node never records these transmission events. In both figures, the blue frame highlights the part of the data content that is associated with a rolling counter, confirming that there is a missing number in

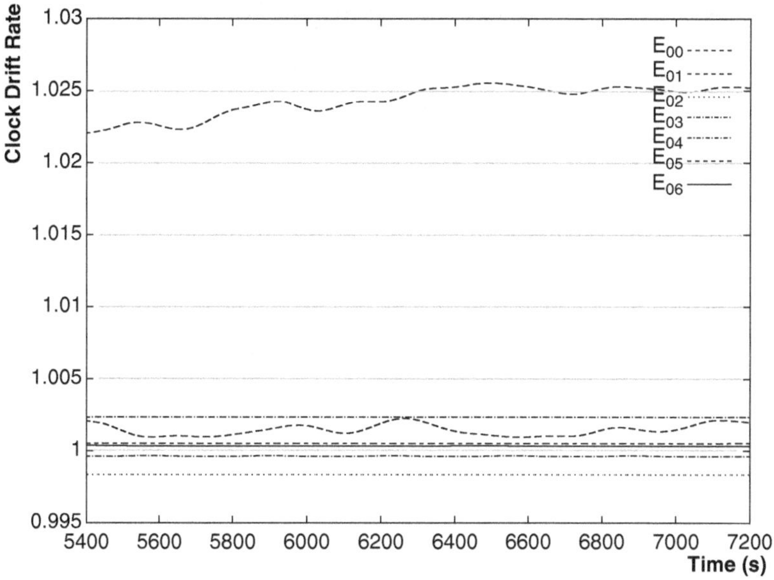

Fig. 7.19 Clock drifts for ECUs on the hybrid vehicle

the sequence. In messages **0x199** and **0x19D**, the first two bits are the signal "Transmission Alive Rolling Count". By observing the trace around the time point when messages are lost, there seems to be enough bus idle time to allow for the transmission of those messages. Hence, we believe that those messages have not been queued at the source node in the first place. The reason behind this behavior may be a filter at the level of the Interaction Layer, but it is difficult to guess without more information on the source node.

7.4.6 Reconstruction of Message-to-ECU Relation in a Hybrid Vehicle

Considering today's automotive supply chain, for IP protection reasons, the system integrator might not provide to the supplier of one node the configuration of all the messages transmitted on the bus by the other nodes. Therefore, in order to recover the information that is necessary for the analysis, the only source of information are the (long) message traces recorded on the bus. In this example, the supplier of an electronic unit for a hybrid vehicle was interested in knowing the worst-case response time for the messages it sent on the CAN bus using the analysis in [22]. However, there was no information available on the other messages transmitted on the bus i.e., information of identifiers, lengths, source nodes and periods were not known. After analyzing a bus trace for a sufficiently long period of time, the bus configuration in Table 7.6 was identified. Apart from the messages transmitted from E_{06} (the only ECU for which knowledge was available), all other messages were easily grouped based on their clock drift (as shown in the table in the *Device* column) and their transmitting ECU was identified. In the process, the true transmission period of each message was calculated. Drift similarities not only occur for the average drift, but also for the deviation of drift over time, as shown in Fig. 7.19, where messages from the same ECU exhibit the same pattern of change of their relative base periods with time.

Chapter 8
CAN Tools

In this chapter, we are going to describe commercial tools for configuring, analyzing and calibrating a CAN communication system. For the sake of generality, we refer to the high level tool flow shown in Fig. 8.1.

The system configuration step pertains to defining the CAN-based networking architecture. Generally speaking, during the system analysis step, the CAN-based networking architecture performance is analyzed with respect to the requirements. For example, a typical scenario that serial data engineers are interested in is to ensure that no critical messages are lost. The system is typically analyzed using either analytical methods and tools (e.g., based upon worst case schedulability analysis) or simulation.

Although the flow is shown as a waterfall process, design iterations may be necessary to tune the system and the protocol stack to achieve the desired performance and cost trade-offs. For example, calibration and testing are typically carried out iteratively. According to the automotive OEM development process standard definition, the term *calibration* typically refers to data that is applied after the software build. Examples of calibration data include, among others, parameters for engine control applications (e.g., RPM thresholds for throttle), variant management (e.g., right-hand side steering wheel variants), etc. In the flow shown in Fig. 8.1 we extend this definition, as calibration is also used to refer to the ability to change the priority, the periodic rate of a message, and to enable/disable the transmission or reception of a message. Each calibration change is then validated through testing. If the testing discovers a discrepancy with respect to the specified timing behavior of the network, then a new calibration is required.

Finally, the flow shown in Fig. 8.1 is only a part of a comprehensive distributed architecture development process, which includes feature development, software development, and system development including the design of the physical components, such as the wiring harness design. The process is used by system and sub-system integrators (e.g., OEM and Tier 1 Suppliers) and pertains to the definition and integration of hardware and software components in a distributed hardware architecture supporting customer features implemented in software.

M. Di Natale et al., *Understanding and Using the Controller Area Network Communication Protocol: Theory and Practice*, DOI 10.1007/978-1-4614-0314-2_8,

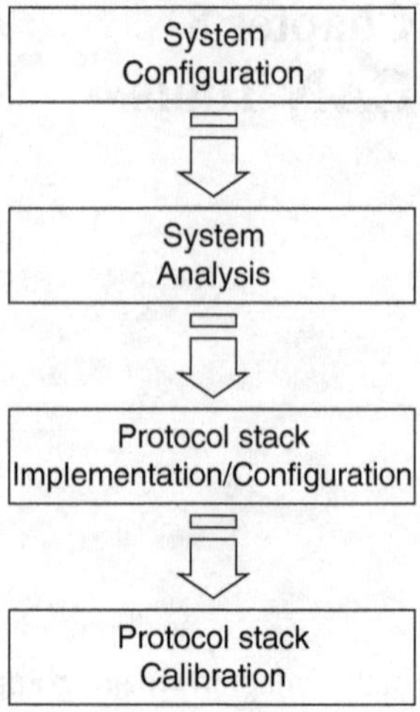

Fig. 8.1 The CAN communication system tool flow

The tools described in this chapter represent a sub-set of the existing tools in the market. We are not aiming at covering the entire competitive landscape. We focus on describing the most relevant features of each tool. Also, we mainly focus on the model-based perspective and only scratch the surface on the interaction between the tools and the CAN adapter hardware or any other hardware emulation device for CAN. The interested reader should refer to each tool data sheet for more details.

8.1 System Configuration

The system configuration step pertains to defining the CAN-based networking architecture, and includes the following:

- The specification of the ECUs (network nodes) that transmit and receive messages through the CAN communication system.
- The specification of the signal interface for each ECU, including the data type (along with a min/max range) of each signal, the mapping of signals into messages with the definition of the start bit in the message, and other attributes of the CAN frame.
- The specification of the signals that an ECU wants to process—as CAN is a broadcast communication system, each signal is sent from the transmitting node to all nodes in the network. Therefore, the designer must specify the signals that each ECU wants to process; all other signals, although contributing to bus traffic and load, are discarded by the ECU.
- The activation policy of each message, which generally includes a periodic mode, an event-driven mode, and possibly a mixed mode. In the periodic mode, each message is transmitted at regular time intervals (plus queuing jitter), and in the

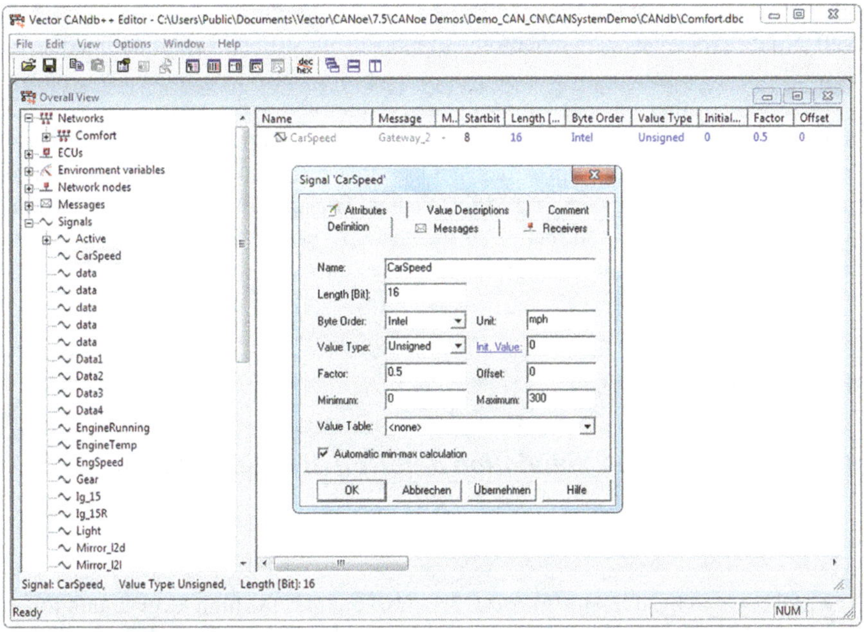

Fig. 8.2 The signal specification in CANdb++

event-driven mode a message is transmitted upon the occurrence of an event (typically a change in the value of one of the signals mapped into the message).

- The identification tag for each message, as each message must have a unique identification tag which is used by the arbitration mechanism.

The tool CANdb++ [12] from Vector Informatik GmbH enables designers to define a detailed specification of the network communication architecture. The tool is part of an overall suite of tools aimed at simulating and monitoring the timing and functional behavior of the network communication architecture. Figure 8.2 shows the tool main window for the specification of a signal carrying a vehicle acceleration data value. On the left hand-side of the figure, the signal browser shows the set of signals in the system. On the right-hand side of the figure, the specific details of the signal are shown, such as the signal length in bits, its type and unit. In addition, other aspects such as the receivers of the signal are displayed.

The tool CANdb++ also enables designers to specify the so-called vehicle program variants. Typically, automotive system integrators define multiple variants of the same network by selecting different configurations. Each variant is a possible alternate implementation of the network. CANdb++ allows designers to define multiple networks in one database. The database represents a superset of all possible real network configurations in one or different vehicles. The tool CANdb++ stores the network database in a proprietary format called "dbc". Although the CAN communication system configuration could be specified using other tools, the format has become the *de-facto* standard for system specification as well as data exchanges with other tools supporting the different steps outlined in the flow.

In addition, other formats are used in the automotive industry to store the CAN communication system configuration, including, among others, Excel and FIBEX. In addition, as AUTOSAR provides services that encapsulate the protocol stacks and related gateway services (e.g., PDU-router, Transport layer, signal router), the AUTOSAR XML-based format is an increasingly popular format that is used to configure protocol specific transport layers (e.g., the CAN transport layer).

8.2 System Analysis

8.2.1 Network Bus Simulation and Rest-Bus Simulation

8.2.1.1 CANoe

The tool CANoe from Vector Informatik GmbH enables the high level simulation of the network timing and functional behavior, and the so-called *rest-bus* simulation. In this scenario, network nodes are simulated using traffic generators defined using the proprietary Vector CAPL language or with ActiveX controls generated from the node definition in the .dbc file. As an option, it is also possible to model the application layer of a node, using a Simulink model. The model outputs and receives signal values, therefore acting as de-facto traffic generator and receiver. Eventually, as the development process proceeds, the virtual ECUs are gradually replaced by real ECUs implemented in hardware as they become available. This means that both the real bus protocol and the simulated protocol co-exist. Once the entire set of virtual ECUs is replaced by a set of real ECUs, the simulated portion of the CAN bus is replaced by the physical bus (Fig. 8.3). In Fig. 8.3, the CANoe software runs on a PC simulating the protocol behavior as well as the nodes transmitting and receiving signals. The real ECUs are connected via the physical bus. The gray and red boxes represent the interface between the physical bus and the PC which enables the communication between any real ECUs and the virtual ECUs.

For CAN, the available statistics are bus load, peak bus load, frames/sec, total frames, errors/sec, total errors, chip state, message frequency, and message counts. The tool supports frame-level simulation. Fault injection is supported by a different tool called CANstress. CANstress can introduce line/line shorts, line/power shorts, and bit corruptions, and impose bus voltage levels and bus resistance levels.

The CAPL programming language provides a complete interface to all frame-level and signal-level network communications. CANoe comes with a variety of pre-programmed functions (CAPL programs, node-level .dlls, etc.) to control more sophisticated interactions with the bus, such as for transport protocols and network management rings.

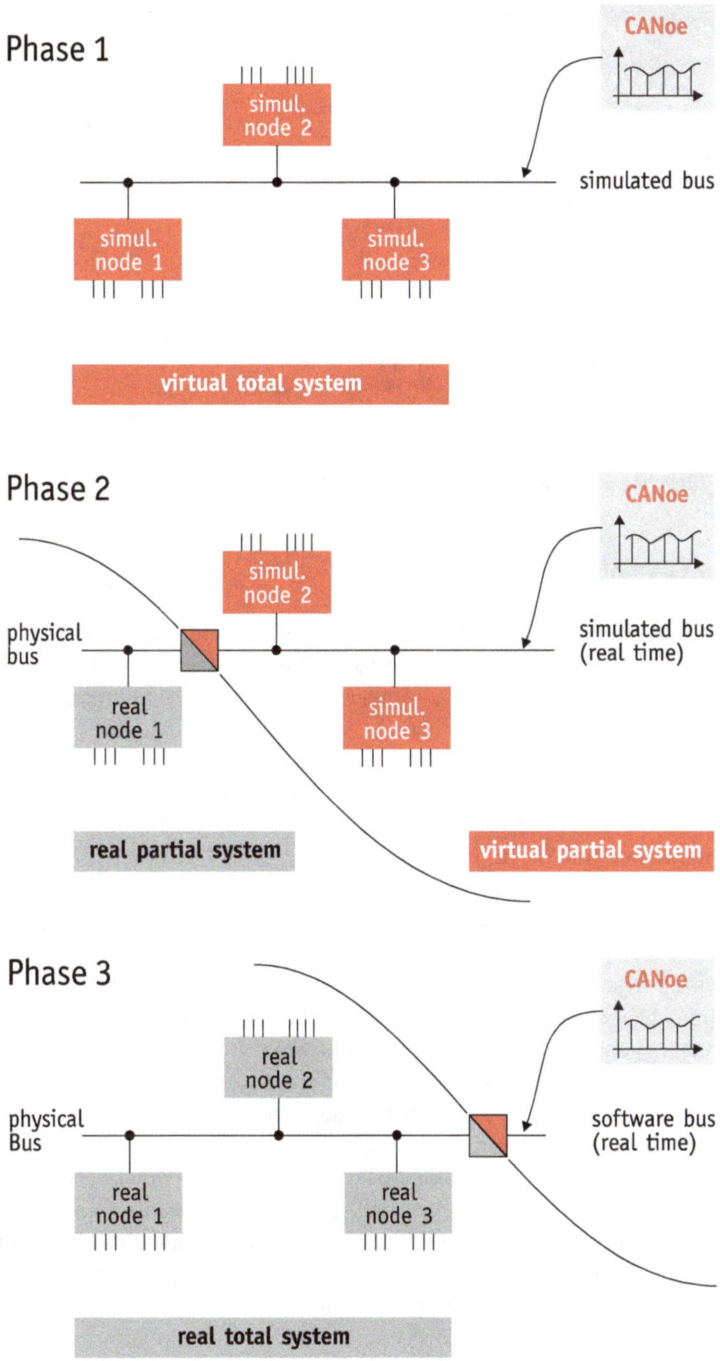

Fig. 8.3 Rest-bus simulation in CANoe (source: Vector Informatik)

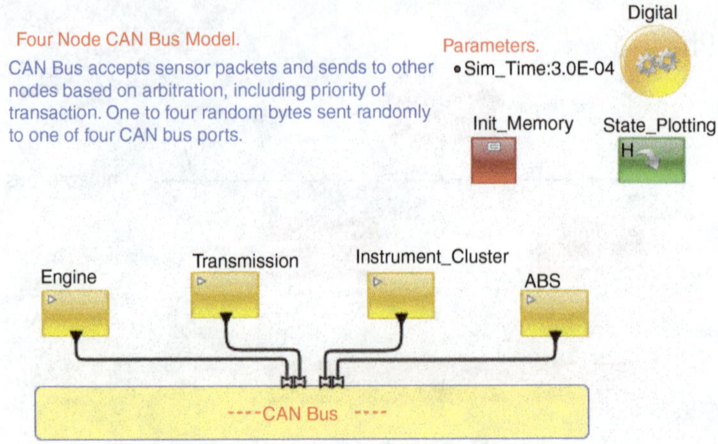

Fig. 8.4 Modeling and simulation of a CAN network in VisualSim (source: Mirabilis Design)

8.2.1.2 VisualSim

The tool VisualSim from Mirabilis Design provides performance analysis, power estimation and architecture exploration capabilities for the design of electronics and real-time software. VisualSim is a graphical modeling and simulation environment that was derived from the academic tool Ptolemy II [24, 3]. Models of the proposed system are constructed in VisualSim using the pre-built parameterized modeling libraries and custom-code blocks using C/C++/Java/SystemC/Verilog/VHDL. The tool provides several different simulation environments based upon the concept of model of computation (MOC). Primary MOCs include Discrete Event, Continuous Time, and Synchronous Data Flow.

The tool enables the simulation of the traffic on the CAN network (Fig. 8.4) using the Discrete Event model of computation where time is modeled by an event calendar that keeps track of the events to be simulated. The assumption is that nothing of interest may happen in between events. Within the tool, modelers can define abstract data structures encapsulating the real data flowing in the network, plus additional fields used to store relevant information used for statistics. For example, one could define a CAN frame data structure to store the time when a frame was placed in the transmit buffer during the simulation. This information can be used at the arriving ECU, during the simulation, to compute the overall transmission latency. The model can contain the description of the network topology, ECU hardware details such as processors and memories, and the application software including the middleware and protocol stack. The model generates a variety of reports including the bus latency, end-to-end segment latency, queue usage, network throughput, and power consumption. The generated reports can be used to size the transmit and receive buffers, and determine the ECU processing capacity requirements and the message priority (ID) assignment. The model can be extended to also study the impact of multiple small packets and the system response to cyber attacks.

CAN simulations are typically performed at the transaction level instead of the bit-level, meaning that the message response times are computed by modeling the high level aspects of the protocol (e.g., the arbitration scheme, the transmit and the receive tasks, etc.) using general purpose resources and schedulers governing the access to the resource or using the Virtual Machine technology described later in this section. As a result, one could model and simulate systems that include random offsets between nodes and clock drifts. In addition, the tool enables the use of data structures that, in addition to modeling the CAN frame format, may include user-defined fields such as the message activation and queuing time. These user defined fields (which do not exist in the real implementation) become handy to debug the model and create useful statistics (e.g., message drop rate). A bit level model and simulation of the CAN network is also possible if the users wants to get a more accurate simulation of the timing performance. Obviously, the more accurate, the slower the simulation.

In the example in Fig. 8.4, the core CAN protocol is modeled using the Virtual Machine technology. This technology is a script and a set of blocks that use a C-like programming format to describe the logic or behaviors. By using a virtual machine code, one can model complex functional and timing behavior, with the advantage of reducing the number of blocks in a diagram. The Virtual Machine accepts data structures on the input ports, processes a sequence of C-like code and outputs a data structure. All the data structures are placed in a single non-priority input queue in the order of arrival. The data structures are executed in the order of arrival. However, it is also possible to model the CAN protocol, using a more traditional approach, as the tool provides modeling artifacts such as priority sorted queues, server resources (to model the bus state, from idle to busy to idle), and routing tables (to define the connectivity between nodes). Finally, both tools provide modeling artifacts to model high abstraction layers in the CAN protocol stack.

Simulation results are displayed using several different primitives including text displays, histograms (Fig. 8.5), Gantt charts, and plots (Fig. 8.7). Figure 8.5 shows an example where, for each master node on the CAN bus, the number of bytes that are transmitted is displayed. Figure 8.6 shows an example where the different states of the bus are displayed including, for example, the transmission of an acknowledgment from the receiving node to the sending node upon the reception of a CAN frame. Figure 8.7 shows the distribution of the message response time (latency) for each node.

8.2.1.3 INCHRON Tool-Suite

The INCHRON Tool-Suite provides modeling, simulation and validation of the dynamic real-time behavior of embedded systems. The tool suite supports the designer in the following tasks:

- Define and verify critical real-time requirements.
- Evaluate system architectures and optimize them regarding timing and performance.

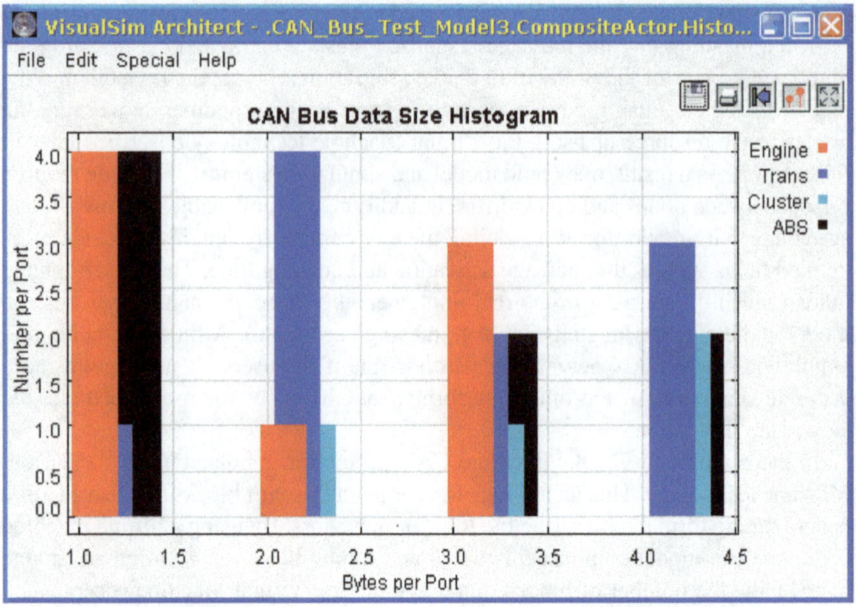

Fig. 8.5 VisualSim histogram of bytes transmitted over CAN Bus (source: Mirabilis Design)

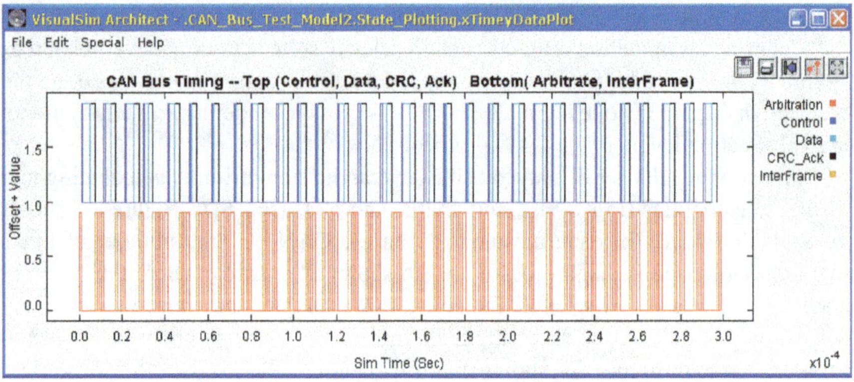

Fig. 8.6 VisualSim plot of the CAN bus states (source: Mirabilis Design)

- Monitor real-time performance and quality in all development phases.
- Enable collaborative design by sharing timing models.

The INCHRON Tool-Suite consists of the real-time simulator chronSIM, the real-time validator chronVAL, and the add-on chronBUS for residual bus simulation and validation. In the following, we will describe the simulation-based tools, chronSIM and chronBUS. In the following section, we will describe chronVAL, the tool for worst/best case schedulability analysis. chronSIM supports the modeling of

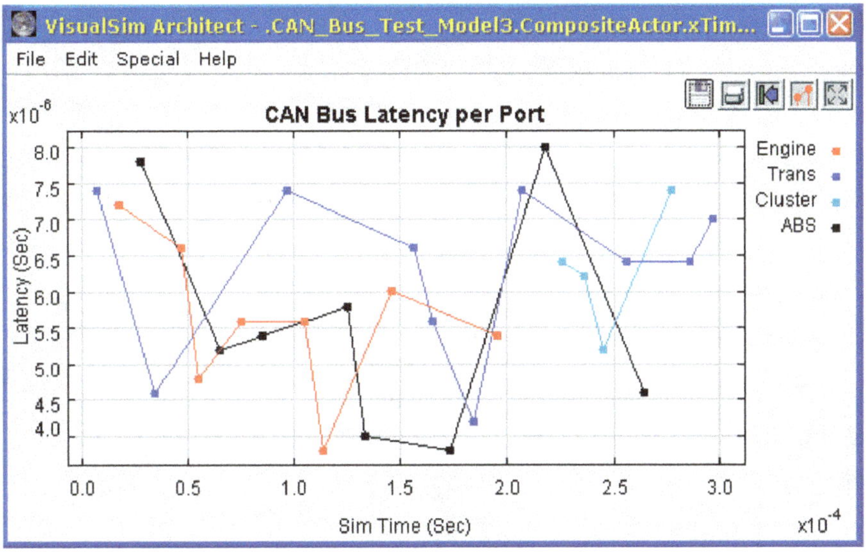

Fig. 8.7 VisualSim plot of CAN bus message latencies (source: Mirabilis Design)

processors and communication buses, clocks, scheduling policies, and input stimuli described with timing models using either the GUI or C code. Figure 8.8 shows the chronVAL GUI. The designer can specify the allocation of tasks and messages to resources (ECUs and CAN buses). For each task or message, the designer can specify the priority (if a priority-based scheduling or arbitration policy is chosen), the activation policy (including the period or cycle time), etc.

The visualization of processes and communications along with the display of real-time requirements enables the designer to analyze the dynamic system behavior, and to discover performance bottlenecks, and integration challenges. The environment enables importing data and models from various formats: Excel csv, UML (profile for Rhapsody), C and C++, and from the PREEvision [13] tool (via the Eclipse EMF). The tool supports HTML reporting and batch processing to enable automated analysis and archiving of results. Figure 8.9 shows an example Gantt Chart. The designer can assess the flow of data across resources (ECUs and buses) to check on interesting scenarios (e.g., data has been lost or oversampled).

chronBUS extends the INCHRON Tool-Suite with residual bus simulation and analysis capabilities. By importing standard communication description files (CANdb or FIBEX [27]), the modeled processors and communication buses can be stimulated with the residual bus communication. The impact on the timing behavior of the residual ECUs resulting from clock offsets and/or drifts can be observed. Each ECU in the network is modeled in detail by the user including the flow of data from the source to the destination ECUs. Alternatively, its model is

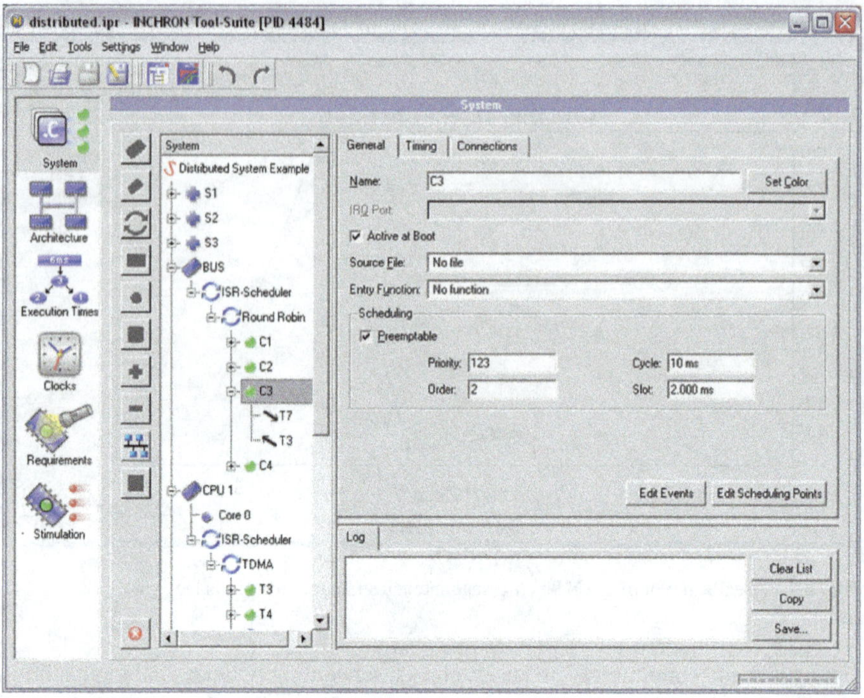

Fig. 8.8 chronVAL GUI (source: INCHRON)

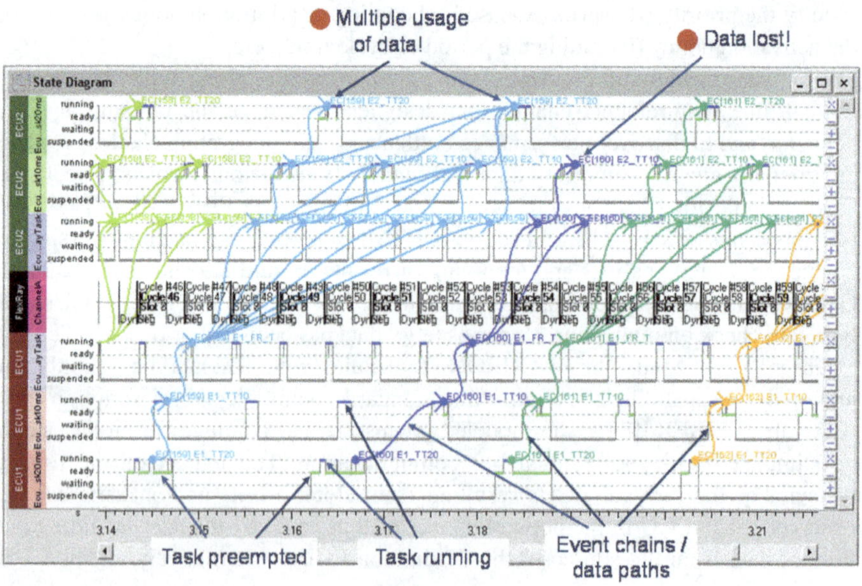

Fig. 8.9 chronVAL Gantt chart (source: INCHRON)

generated by chronBUS automatically. An automatically generated analysis report in HTML format provides detailed information for each CAN id including statistics and histograms.

chronSIM models peripherals as execution units that have an independent and well specified behavior. Register accesses of the simulated machine code prepare CAN messages and flag them as ready for transmission. If the bus is free, the peripheral model starts an arbitration, otherwise it waits until the current message transmission finishes. All messages available in the peripherals connected to the bus are considered during the emulation of the CAN arbitration scheme. Due to performance reasons, the bus is not modeled at the bit level. Instead of comparing recessive with dominant bits, CAN identifiers as a whole are compared. In the discrete event based simulation of chronSIM, the comparison occurs at a point of time that is about a sampling interval after the beginning of the transmission of the earliest CAN message. During this time, all peripheral models are allowed to add their message to the arbitration. Afterwards, the bus is considered busy as at least one message has been granted access to the bus and all other ECUs switch to a listening mode. The next simulation event for the CAN bus is scheduled at the end of the message which has won the arbitration. This point in time is calculated from the true length of the message, including its specific data content and the resulting bit stuffing. Then, the ACK bit created by the receivers is simulated and the next event is scheduled after the interframe space. Because of the chronSIM's multi-resource simulation capabilities, all the simulation events for all CAN buses and their peripherals are processed and interleaved with the simulation events from the software tasks running on the simulated microprocessors and their other peripherals. This simulates the real parallel timing behavior on multiple resources (ECUs and buses). In addition to CAN buses, chronSIM also supports FlexRay and full-duplex switched Ethernet systems. For pure CAN-based systems, the chronBUS extension simulates the behavior of CAN buses without explicitly modeled processors and software tasks. Message activations are modeled by traffic generators. Figure 8.10 shows the chronBUS GUI. The designer can use it to specify the set of ECUs connected to the CAN bus and the messages transmitted over the CAN bus. The design can also be imported from CAN-DB or FIBEX files.

Figure 8.11 shows the chronBUS Gantt chart. For example, the timeline at the bottom of Fig. 8.11 shows the different states of the CAN bus over time, the time duration of each state, as well as the ID of the message being transmitted. This information can be used to debug the overall communication scheme and identify timing bottlenecks (e.g., message losses).

8.2.2 *Traffic Logging and Analysis*

The tool CANalyzer by Vector Informatik GmbH enables the real-time logging of the traffic (Fig. 8.12) on the CAN network. In addition, the user can also transmit additional data traffic or reply to previously observed messages. In Fig. 8.12, an example of a trace is shown (right-hand side of the GUI).

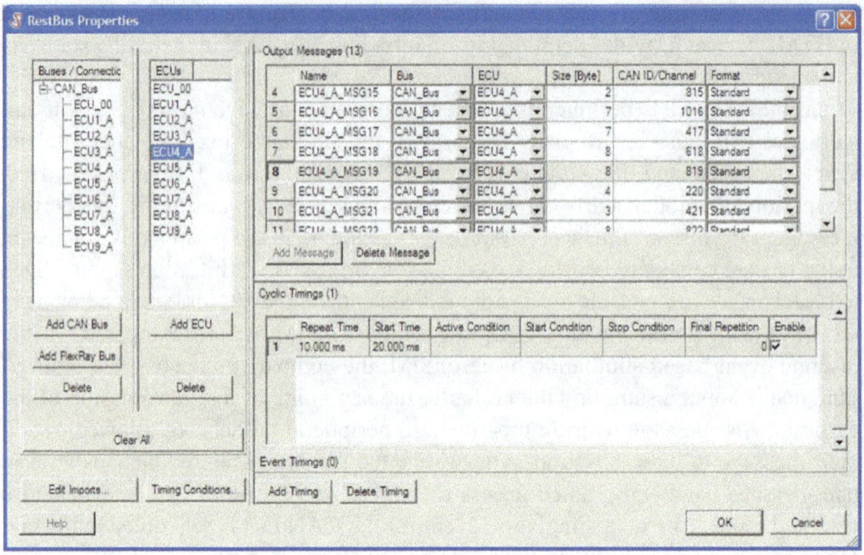

Fig. 8.10 chronBUS GUI (source: INCHRON)

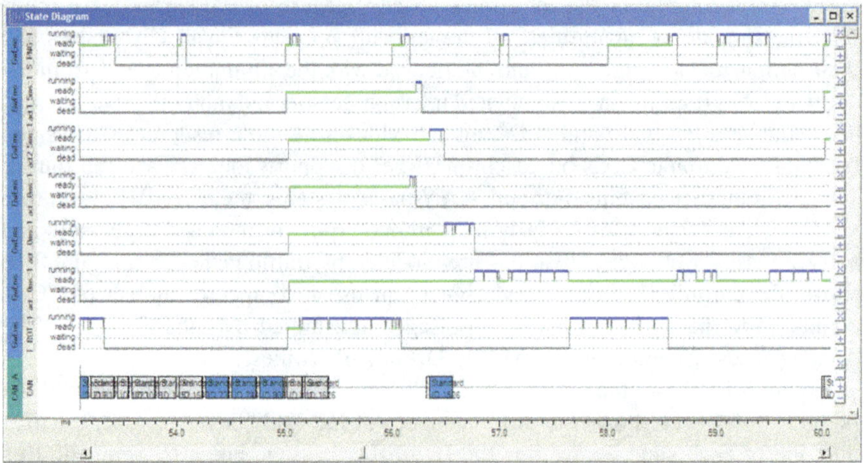

Fig. 8.11 chronBUS Gantt chart (source: INCHRON)

The traffic is typically stored as trace data that includes the CAN id, the arrival times of the frames, and signal values. A more detailed example of the traffic log is shown in Fig. 8.13. Generally speaking, the trace logged data include information such as the timestamp, channel number (if working with multiple busses, channels), frame ID, frame name, Rx/Tx direction, frame DLC, signal data—raw encoded values and decoded symbolic values, the data length (for the payload part), and the data content.

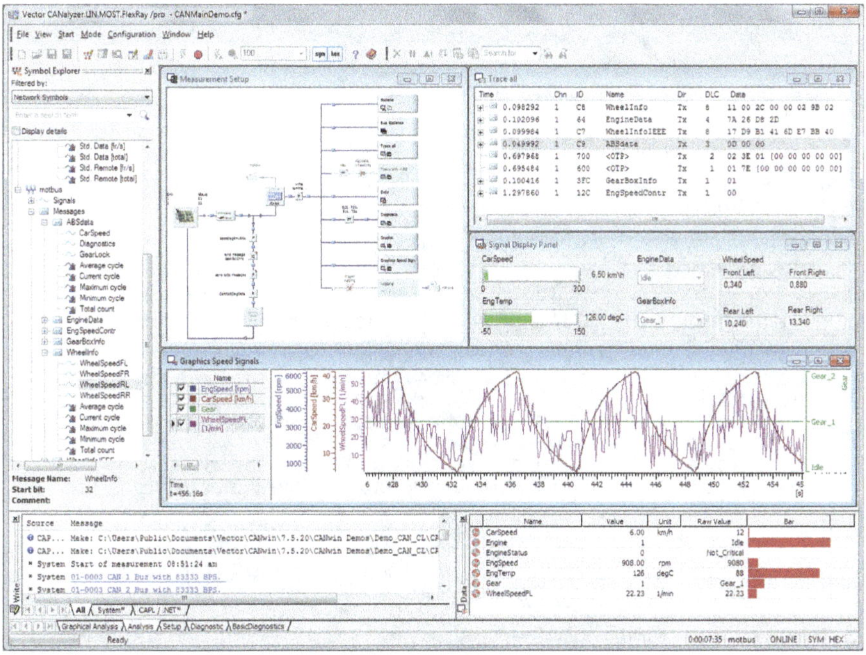

Fig. 8.12 CANalyzer configuration and analysis (source: Vector Informatik)

8.2.3 Network Bus Worst/Best Case Analysis

The SymTA/S tool suite from Symtavision performs automatic real-time analysis of the network communication. SymTA/S analyzes details of the temporal communication behavior including the bus arbitration (scheduling), and computes related timing properties such as:

- Bus load (average and peak load)
- Cycle time deviations and message jitter
- Total message delay including sender-side arbitration delay
- Burst lengths (bus busy times)

For fixed-priority scheduled systems, the required attributes of a message for network analysis are: the period (if the message is periodically activated) or minimum inter-arrival times, the worst case transmission time (or the number of bytes in the payload), its priority, and (when known) its queuing jitter.

These properties are valuable in the validation of the acceptance criteria of bus configurations. Specifically, network timing analysis enables the validation and optimization of network configurations in two scenarios:

Late-stage analysis to support system test and validation: Analysis is based on existing bus traces, e.g. from CANoe and CANalyzer. The Symtavision tool

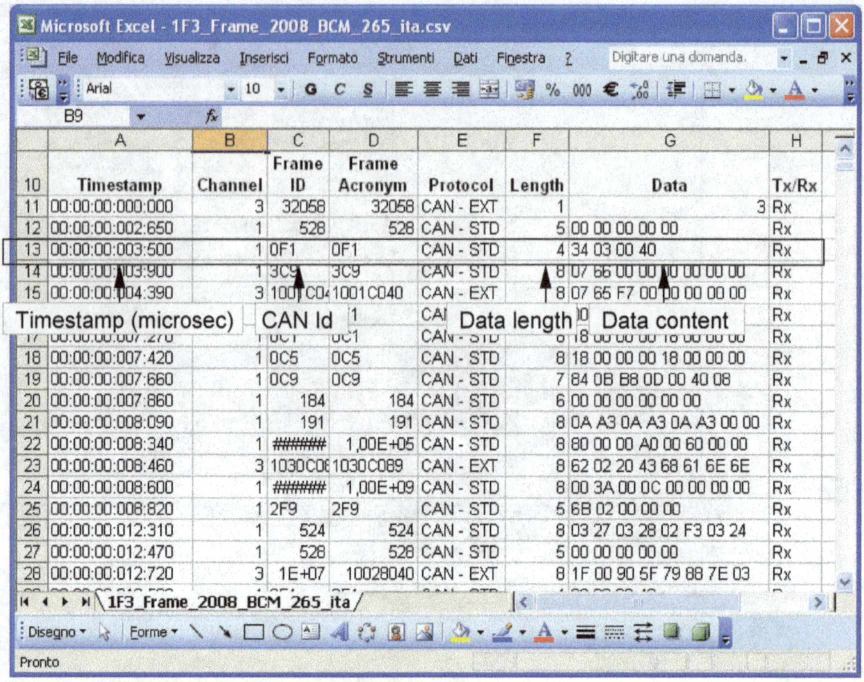

	A	B	C	D	E	F	G	H
10	Timestamp	Channel	Frame ID	Frame Acronym	Protocol	Length	Data	Tx/Rx
11	00:00:00:000:000	3	32058	32058	CAN - EXT	1	3	Rx
12	00:00:00:002:650	1	528	528	CAN - STD	5	00 00 00 00 00	Rx
13	00:00:00:003:500	1	0F1	0F1	CAN - STD	4	34 03 00 40	Rx
14	00:00:00:003:900	1	3C9	3C9	CAN - STD	8	07 66 00 00 00 00 00 00	Rx
15	00:00:00:004:390	3	1001C04	1001C040	CAN - EXT	8	07 65 F7 00 00 00 00 00	Rx
16	[illegible]	[illegible]	[illegible]	[illegible]	[illegible]	[illegible]	[illegible]	Rx
17	[illegible]	1	[illegible]	[illegible]	[illegible]	[illegible]	[illegible]	Rx
18	00:00:00:007:420	1	0C5	0C5	CAN - STD	8	18 00 00 00 18 00 00 00	Rx
19	00:00:00:007:660	1	0C9	0C9	CAN - STD	7	84 0B B8 0D 00 40 08	Rx
20	00:00:00:007:860	1	184	184	CAN - STD	6	00 00 00 00 00 00	Rx
21	00:00:00:008:090	1	191	191	CAN - STD	8	0A A3 0A A3 0A A3 00 00	Rx
22	00:00:00:008:340	1	#######	1,00E+05	CAN - STD	8	80 00 00 A0 00 60 00 00	Rx
23	00:00:00:008:460	3	1030C08	1030C089	CAN - EXT	8	62 02 20 43 68 61 6E 6E	Rx
24	00:00:00:008:600	1	#######	1,00E+09	CAN - STD	8	00 3A 00 0C 00 00 00 00	Rx
25	00:00:00:008:820	1	2F9	2F9	CAN - STD	5	6B 02 00 00 00	Rx
26	00:00:00:012:310	1	524	524	CAN - STD	8	03 27 03 28 02 F3 03 24	Rx
27	00:00:00:012:470	1	528	528	CAN - STD	5	00 00 00 00 00	Rx
28	00:00:00:012:720	3	1E+07	10028040	CAN - EXT	8	1F 00 90 5F 79 88 7E 03	Rx

Fig. 8.13 Traffic data log (source: Vector Informatik)

TraceAnalyzer imports CANoe bus traces and extracts the timing properties (e.g., message jitter) at the bus, channel, and message level. Once extracted, the designer can validate the temporal behavior of the system against the desired performance figures (e.g., maximum bus load less than 40%, maximum jitter less than 0.5ms, etc). The validation can cover system normal operation, as well as start-up, diagnosis, flashing, etc. Optionally, the user can subsequently perform a worst-case bus traffic analysis to increase the reliability of the gained acceptance criteria.

Early-stage forecast as part of bus configuration and network design: the analysis is based on configuration data imported from existing data bases (DBC, FIBEX, AUTOSAR, etc.) or directly captured within the Symtavision tool SymTA/S Network Designer. The tool is able to import the bus configuration and transform it into a bus timing model, which is then used for timing analysis. The analysis is based on mathematical calculations that have their roots in the scheduling analysis theory. SymTA/S also extended the analysis capabilities to include statistical evaluations in addition to worst and best case scenarios.

As the tool enables the designer to predict (under the assumptions of real time scheduling theory) the message response time and jitter, this information can be

Table 8.1 Characteristic functions of most relevant event streams

Model	Parameters	$\eta^+(t)$	$\delta^-(n)$
Periodic	$<P>$	$\lceil \frac{t}{P} \rceil$	$(n-1) \times P$
Periodic w/jitter	$<P, J>$	$\lceil \frac{t+J}{P} \rceil$	$\max\{0, (n-1) \times P - J\}$
Periodic w/burst	$<P, J, d>$	$\min\left(\lceil \frac{t+J}{P} \rceil, \lceil \frac{t}{d} \rceil\right)$	$\max\{0, (n-1) \times P - J, (n-1) \times d\}$

used to optimize the network utilization, by modifying the message offsets. By using offsets, the user can balance the communication and create *gaps in the schedule* that provide room for other traffic.

SymTA/S is Eclipse-based and provides an environment for importing and editing network configurations. A tree view captures the network structure and spreadsheet editors enable parameter control. Copy&paste functionality to and from Excel is provided as well as other formats including XML, CSV, etc. Finally, PDF reports are generated automatically, and the integrated Python scripting engine enables integration with other tools and user customizable add-ons. Optional GUI views provide additional visualizations such as network topology view, signal sequence diagrams, etc.

SymTA/S uses the notation of the activating event streams [57]. Instead of relying strictly on periods (possibly with jitter), SymTA/S accepts any arbitrary event stream, as long as the following two characteristic functions are supplied:

- $\eta^+(t)$: the maximum number of events within a given interval of time t
- $\delta^-(n)$: the minimum distance between n events

For example, by substituting the two functions in (3.5), the equations to compute the worst case queuing delay and response time are as follows:

$$w_i(q) = (q+1) \cdot C_i + B_i + \sum_{\forall \tau_j \in hp(i)} \eta^+(w_i(q)) \cdot C_j \tag{8.1}$$

$$R_i^{\max} = \max_{q=1,2,\ldots} (w_i(q) - \delta^-(q+1)) \tag{8.2}$$

These characteristic functions can be adjusted to a variety of event streams, including periodic, jitter, burst, etc. Table 8.1 summarizes the most relevant such functions.

The tool computes both the worst and the best case response times for CAN messages using either 11 or 29-bit long IDs. In addition, the tool computes jitters at the receiving side and the bus utilization in the worst and the best case scenarios.

The tool GUI is illustrated in Fig. 8.14, while Fig. 8.15 shows the analysis results for one message, in this case the message with identifier `0x0F2` from ECU1. In the top left side of the GUI, we see the system structure tree and next to it the load overview. For each message, the upper middle of the GUI show the minimum and maximum total latencies (response times) including a deadline marker, while the upper right side of the GUI shows the statistical distribution results. The message

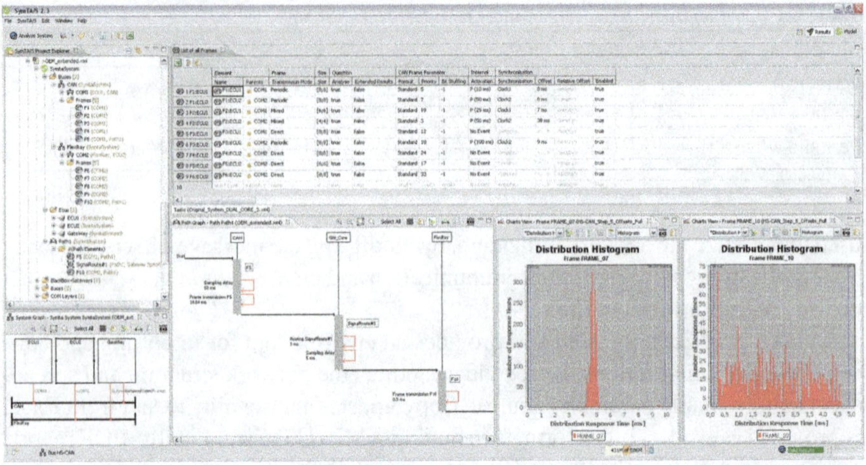

Fig. 8.14 SymTA/S bus timing model

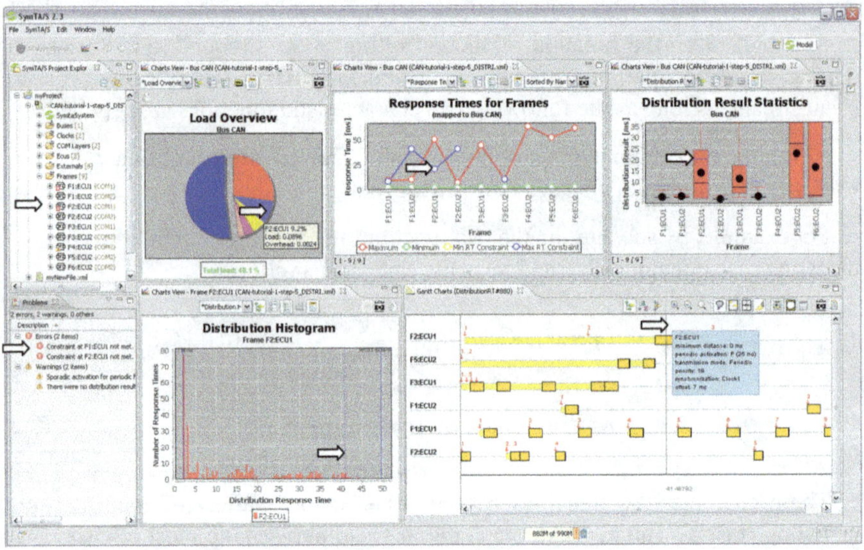

Fig. 8.15 SymTA/S timing analysis and forecast results

`0x0F2` from ECU1 violates its deadline because of higher-priority CAN traffic as shown in the Gantt chart on the bottom right part of the figure, displaying the computed worst-case scenario for the transmission of the message. The statistical analysis (top-right side of the window) shows that the user should expect a deadline violation on average in 8.7% of all cases.

chronVAL from INCHRON provides modeling, analysis and validation of the real time behavior of networked embedded systems. The formal mathematical

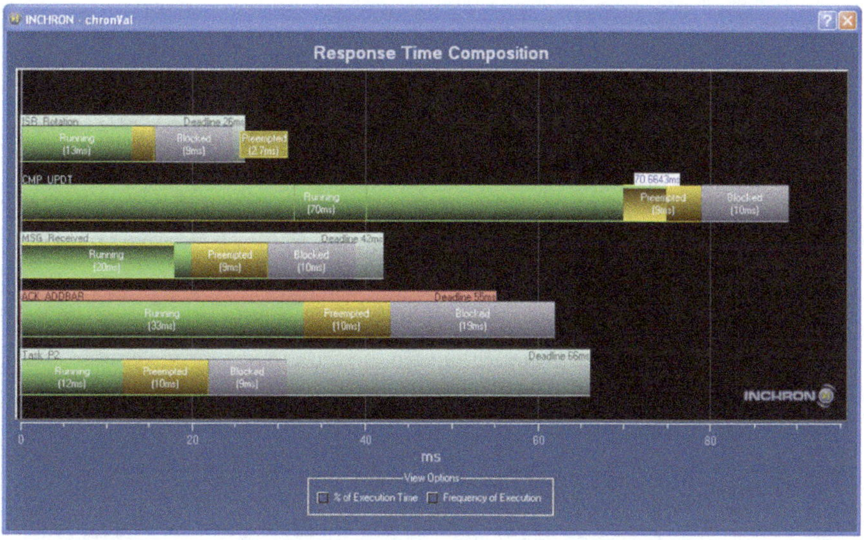

Fig. 8.16 chronVAL Gantt chart (source: INCHRON)

approach finds the best and worst case stimuli that lead to the best and worst case response times of the system including end-to-end communication over multiple processors and buses. chronVAL displays detailed system information about maximum CPU and bus loads, burst lengths as well as task blocking and suspension times. The integration with the real-time simulator chronSIM allows to combine the advantages of simulation and validation unfolding in a real-time view to the system.

chronVAL models messages and their connection to the sending peripheral explicitly. The transmission of messages occurs either periodically or event-based. Messages can be part of a larger system and therefore be connected to sending or receiving tasks or interrupts and may be part of event chains. chronVAL calculates the guaranteed worst-case and best-case transmission times for the messages using the message parameters, such as the offset, the bus load and speed, and taking into account the asynchronous clocks of different peripherals. It uses methods from schedulability analysis and the real-time calculus providing detailed interval-based diagrams for bus load and messages stimulations. The tool has several visualization aids that help the designer assess the performance of the system. In Fig. 8.16 another example of Gantt chart is shown. Both messages and tasks (including interrupt service routines) are displayed in the Gantt chart, including a breakdown of their response times (running time, preempted time, and blocked time).

In Fig. 8.17 the simulation and worst/best case analyses are combined in a single table (e.g., average and worst/best case numbers for arbitration and message latencies are displayed).

Figure 8.18 shows the same results of Fig. 8.17 at the message level (e.g., the bus load, the arbitration and the latency delays associated to a specific message).

Id	Type	Count	Avg Length	Load	Arb Avg	Arb Min	Arb Max	Arb Total	Lat Avg	Lat Min	Lat Max	Lat Total
113 (0x71)	Standard	60407	605.999 us	6.06 %	96.306 us	0 ps	4.451 ms	5.817600 s	702.296 us	600.000 us	5.057 ms	42.423653 s
205 (0xcd)	Standard	1209	324.000 us	0.06 %	136.965 us	0 ps	1.133 ms	165.591 ms	460.959 us	318.164 us	1.457 ms	557.300 ms
226 (0xe2)	Standard	6043	378.000 us	0.38 %	140.080 us	0 ps	1.512 ms	846.506 ms	518.066 us	372.026 us	1.890 ms	3.130673 s
233 (0xe9)	Standard	1209	546.000 us	0.11 %	144.564 us	0 ps	1.357 ms	174.778 ms	690.548 us	540.777 us	1.903 ms	834.873 ms
335 (0x14f)	Standard	3023	558.000 us	0.28 %	270.650 us	0 ps	1.926 ms	818.176 ms	828.643 us	552.574 us	2.484 ms	2.504988 s
349 (0x15d)	Standard	1210	384.000 us	0.08 %	91.244 us	0 ps	1.448 ms	110.406 ms	475.237 us	380.314 us	1.832 ms	575.037 ms
414 (0x19e)	Standard	1209	492.000 us	0.10 %	718.626 us	619.000 us	2.497 ms	868.819 ms	1.210 ms	1.111 ms	2.989 ms	1.463647 s
502 (0x1f6)	Standard	1209	612.000 us	0.12 %	517.277 us	336.164 us	3.409 ms	625.388 ms	1.129 ms	948.164 us	4.021 ms	1.365296 s
510 (0x1fe)	Standard	12081	672.000 us	1.34 %	242.976 us	0 ps	4.039 ms	2.935403 s	914.965 us	666.062 us	4.711 ms	11.053702 s
522 (0x20a)	Standard	6043	678.000 us	0.68 %	582.811 us	390.026 us	4.729 ms	3.521927 s	1.260 ms	1.068 ms	5.407 ms	7.619081 s
523 (0x20b)	Standard	3022	606.000 us	0.30 %	1.345 ms	1.086 ms	5.425 ms	4.067351 s	1.951 ms	1.692 ms	6.031 ms	5.898683 s
527 (0x20f)	Standard	2417	612.000 us	0.24 %	920.404 us	0 ps	6.049 ms	2.224618 s	1.532 ms	606.893 us	6.661 ms	3.703800 s
606 (0x25e)	Standard	605	492.000 us	0.05 %	1.988 ms	1.660 ms	6.679 ms	1.203144 s	2.480 ms	2.152 ms	7.171 ms	1.500804 s
607 (0x25f)	Standard	30201	612.000 us	3.06 %	349.299 us	0 ps	7.189 ms	10.549198 s	961.289 us	606.103 us	7.801 ms	29.031917 s
615 (0x267)	Standard	2417	492.000 us	0.20 %	1.071 ms	618.601 us	7.819 ms	2.588972 s	1.563 ms	1.110 ms	8.311 ms	3.778136 s
631 (0x277)	Standard	605	384.000 us	0.04 %	1.493 ms	1.134 ms	8.334 ms	903.281 ms	1.877 ms	1.518 ms	8.718 ms	1.135601 s

Fig. 8.17 chronBUS table of results (source: INCHRON)

CAN message id 527 (0x20f) Standard on bus CAN_0

Arbitration Times		Latency Times	
Average	920.404 us	Average	1.532 ms
Minimum	0 ps @ 3.248918 s	Minimum	606.893 us @ 445.101781 s
Maximum	6.049 ms @ 0 ps	Maximum	6.661 ms @ 0 ps
Total	2.224618 s	Total	3.703800 s

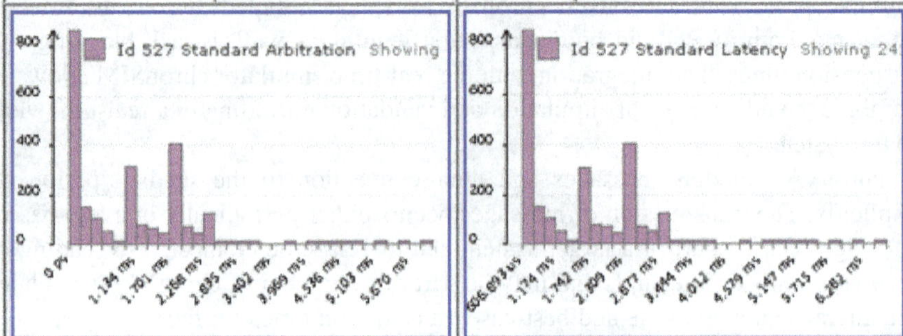

Count	2417
Message Length	612.000 us
Load	0.24 %

Fig. 8.18 chronBUS single message results (source: INCHRON)

Figure 8.19 shows the event spectrum viewer and the results of the formal analysis, depicting in an event spectrum the resource load over growing time intervals. The maximum load (max. requested computation time) of the 5 ms task in any 10 ms interval is, in the worst case, equal to 3.9 ms (three activations with a net execution time of 1.3 ms each). The maximum remaining capacity available to

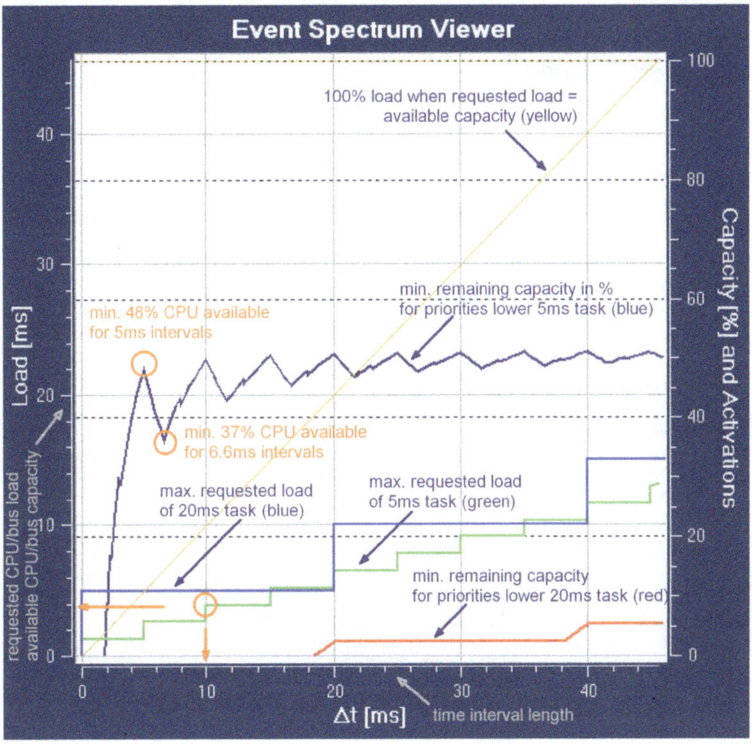

Fig. 8.19 chronBUS event spectrum viewer (source: INCHRON)

processes with a priority lower than the 5 ms task is shown in percentage. For large intervals the remaining capacity is about 50%, but for shorter intervals the capacity varies. If an additional functionality needs to be activated in an interval of length 6.6 ms, the guaranteed capacity would be even lower than 5 ms. System architects use this interactive diagram to perform fast design space explorations. Other results shown are the maximum number of activations for a process.

8.3 Protocol Stack Implementation and Configuration

8.3.1 CAN Driver Development and Integration

Vector Informatik GmbH provides basic communication-related software called CANbedded Basic Software that enables ECUs to exchange information (data, diagnostic, etc.) over the CAN bus. CANbedded consists of configurable software

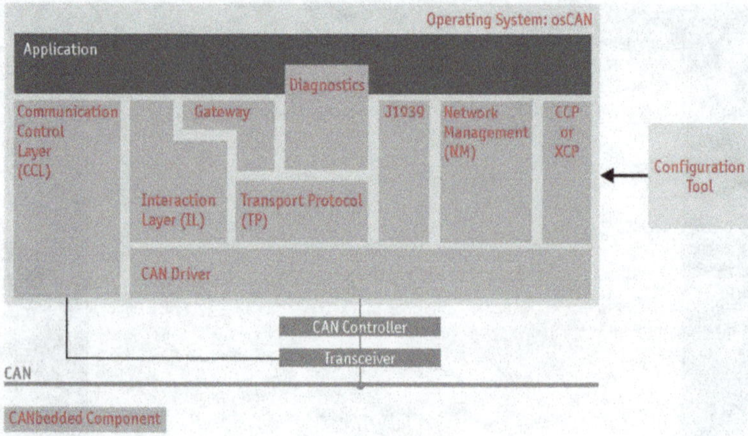

Fig. 8.20 η^{+}-CANbedded basic software

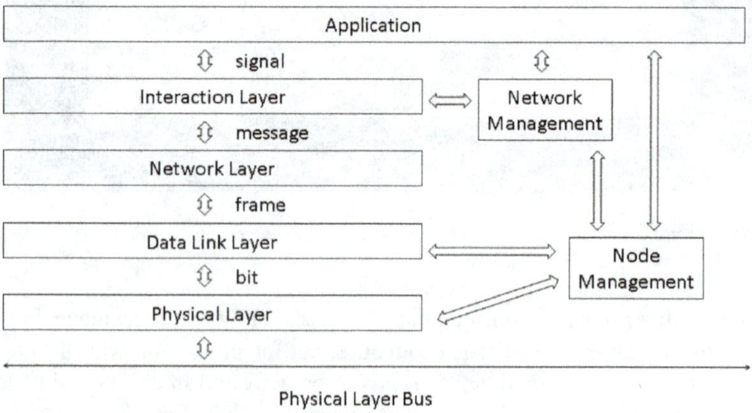

Fig. 8.21 η^{+}-CANbedded basic software abstraction layers

components including the CAN driver, Interaction Layer, Network Management, Transport Protocol, and Communication Control Layer (Fig. 8.20). The CANbedded software stack handles OEM-conformant sending and receiving of messages over CAN (Fig. 8.21).

CANbedded is provided in two variants, one for single-channel communication and one for multiple CAN channels. CANbedded can be extended with additional basic software from Vector, for example, for measurement and calibration (CCP—see Sect. 9.3).

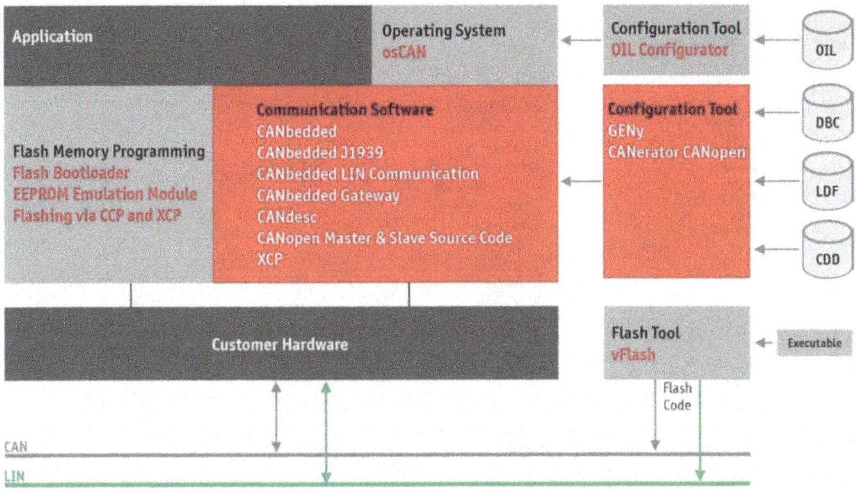

Fig. 8.22 Vector configuration tools for CAN

The interaction layer provides an implementation of the standard features for this layer, as described in Chap. 4, separating the basic driver (depending on the data link layer) and the actual application (independent of the basic bus system). The Vector interaction layer has the following features:

- Signal-oriented interface
- Sending CAN messages corresponding to the specified transfer types
- Monitoring of incoming messages
- Different notification types

The CAN Driver provides a hardware independent (as much as possible) interface to higher software layers. All necessary settings such as parameters for the hardware acceptance filter or Bus Timing Register are made at configuration time. The configuration tool GENy (formerly called CANGen) from Vector Informatik GmbH offers configuration options for these filters and registers. The tool is part of a set of tools provided by Vector Informatik GmbH aimed at providing configuration capabilities for all layers of the CAN protocol stack (Fig. 8.22). The interaction layer provides a signal oriented interface including special features that are specific for vehicle manufacturers.

The tool GENy takes as input a Vector DBC file. One of the main features of the GENy tool is the ability to control the generation of the interaction layer APIs that are used by the application. Finally, the GENy tool improves the CANGen tool as it provides an XML-based data storage and configuration, in addition to the configuration capabilities available from the GUI (Fig. 8.23).

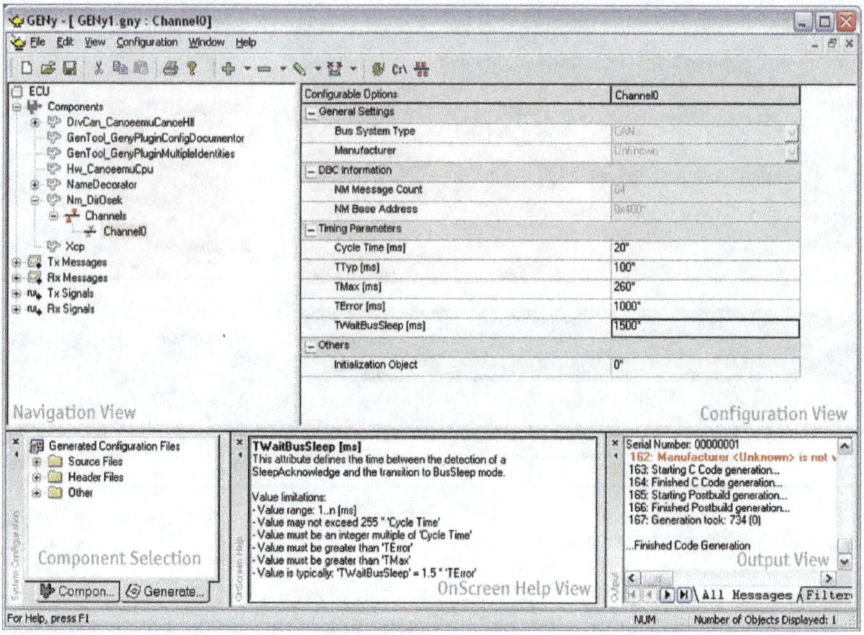

Fig. 8.23 GENy configuration tool

8.4 Protocol Stack/Application Calibration

Developed and by Ingenieurb uro Helmut Kleinknecht, a manufacturer of calibration systems, the CAN Calibration Protocol (CCP) is used across several applications in the automotive industry. Eventually, CCP became an ASAP[1] standard and additional features were added.

Generally speaking, CCP can be used in several stages of the in-vehicle network development cycle, including, among others, the development of electronic control units (ECU), functional and environmental tests of an ECU to test systems and stands for the controlled devices (e.g., combustion engines, gearboxes, suspension systems, etc.), on-board test and measurement systems of pre-series vehicles, and any non-automotive application of CAN-based distributed electronic control systems.

The CAN Calibration Protocol is essentially a software interface used to interconnect a development tool with an Electronic Control Unit (ECU). The interface defines methods to handle module calibration, measurement data acquisition,

[1]The ASAP task force (Arbeitskreis zur Standardisierung von Applikationssystemen; English translation: Standardization of Application/Calibration Systems task force) has been founded by the companies Audi AG, BMW AG, Mercedes-Benz AG, Porsche AG and Volkswagen AG.

and flash programming activities. CCP is strictly for the CAN physical layer. XCP is the more recent extension of CCP that is truly independent of the physical layer and can be operated on top of several networks, including CAN, Ethernet, and FlexRay. CCP is capable of supporting a single point-to-point connection or a networked connection to a larger distributed system. This means that any combination of calibration, measurement data acquisition, and flash programming activities are possible for a single module or any number of modules across a CAN network.

CCP defines the communication of controllers with a master device using the CAN 2.0B (11-bit and 29-bit identifier), which includes 2.0A (11-bit identifier) for the data acquisition from the controllers, and the memory transfers to and from control functions in the controllers for calibration.

The CAN Calibration Protocol is used as a monitor program. Similar to many earlier RS232-type monitors and bootstrap loaders that provide basic read and write memory capabilities (e.g., for RAM, PORTS, ROM, and FLASH), CCP provides the same functionality using a standard protocol rather than a company-specific proprietary protocol. When a high-speed CAN bus is used, CCP, unlike some previous 9,600 baud UART-based monitors, provides the ability to access data at a fast rate while the application is running on the target node. CCP uses a specific conversation or dialog to accomplish each designated function with only two CAN identifiers for message transfers. Most CCP dialogs use a Master/Slave form of conversation. Each dialog is a collection of messages exchanged between the Master, usually the calibration or development tool, and a Slave ECU. In the CCP dialogs provided by most monitor programs, the PC-based monitoring tool functions as the Master. The tool (or Master) always initiates the conversation with a single CAN message. Once this message has been received, the ECU (or Slave) is then responsible for responding with a single CAN message. ECUs do not send information without the Master (tool) initiating commands.

Basic development uses for CCP include:

- Real-time information from the ECU (basic read and write functions)
- Real-time access to ECU parameters (data acquisition)
- Real-time adjustment of the ECU's process algorithms (Calibration)
- In-system or in-vehicle evaluation of design concepts
- Evaluation of engineering design modifications
- In-system (or in-vehicle) Flash Programming
- Emulation-type operation beyond the lab bench

Further details on CCP can be found in Chap. 9, dedicated to the higher-level protocols based on CAN.

Chapter 9
Higher-Level Protocols

The CAN standard defines the lower two layers of the ISO reference model, and even those with several gaps and omissions. This is often not enough for the practical operation of a network and the devices attached to it. Interchangeability and portability of devices require further standardization of cables and connectors at the physical layer. Transmission of data units longer than 8 bytes needs to be supported by transport-level protocols. Furthermore, several classes of users need to agree on additional higher-level protocol features, including data types, type encoding, and the definition of commonly used parameters and parameter groups. Also, network management features, addressing of multicast groups and definition of multicast addresses may be subject to standardization. It is not surprising, then, that a number of additional standards have been developed to provide solutions with respect to these problems. These standards typically span across all levels of the protocol stack. They provide specifications for bus lengths and bit rates as well as connectors and cable types at the physical layer. They include a transport-level protocol and a network management protocol. Very often, they also provide standardization of types and parameters and an addressing scheme overlaid on top of CAN identifiers.

9.1 J1939

J1939 is a standard issued by the Society of Automotive Engineers (SAE), with its Truck and Bus Control and Communications subcommittee. It was initially aimed at truck trailers and other heavy-duty vehicle communication, but has become increasingly popular in other domains, including marine systems, plant and factory control. The J1939 specification is not freely available. It can be purchased from SAE as a collection of standards or as a single pdf document. Even then, the designer needs to summon all of his/her patience to go through the 1,600 pages

M. Di Natale et al., *Understanding and Using the Controller Area Network Communication Protocol: Theory and Practice*, DOI 10.1007/978-1-4614-0314-2_9,

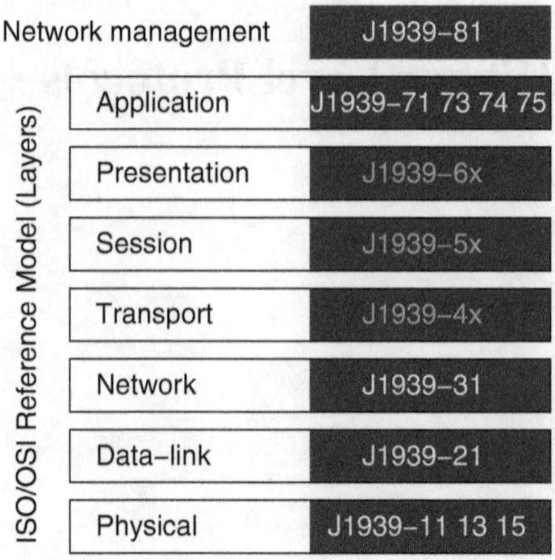

Fig. 9.1 J1939 standards in the ISO/OSI reference model

of the complete specification. This section is clearly only a short tutorial, aimed at covering the main protocol concepts. It draws from a number of sources listed in the bibliography. For a more detailed description, a possible starting point is [62].

J1939 defines communication at different levels of the communication stack, from the physical layer, to the transport protocol, network management and the standardization of message content through the definition of a set of *Parameter groups*. The J1939 Standards Collection was designed to follow the ISO/OSI 7-Layer Reference Model, addressing each layer by a corresponding set of documents (Fig. 9.1). At the time this book was written, only some layer specifications had been defined. Only the documents marked in white in Fig. 9.1 are currently available.

Part of the protocol characteristics is defined by the requirement to provide (at least partial) backward-compatibility with the functions of the older RS485-based communication protocols (J1708/J1587).

Characteristics of the CAN-based J1939 protocol include the use of 29-bit identifier messages and the option of point-to-point or broadcast communication. With respect to similar protocols, J1939 is the only one to request extended identifiers and does not require the definition of a master. Extended CAN identifiers are actually the result of a request from SAE, which needed additional addressing capability.

In J1939, each CAN node is referred to as an Electronic Control Unit (ECU). Each ECU has at least one *Name* and one node *Address*, but it may also have more than one in the same application or network configuration. J1939 ECU names are 64-bit labels, and node addresses are 8-bit identifiers.

9.1.1 Physical Layer

As most other higher-level protocols, J1939 is not limited to the introduction of missing features at the transport and application layers, but goes back to the physical

Table 9.1 Physical layer requirements for J1939

Item	Requirement
Line driver	Differential bus (2 lines)
Network access arbitration	Random, nondestructive, bit wide
Baud rate	250,000
Maximum nodes	30
Topology	Linear bus
Trunk length	40 m (120 ft.)
Drop length	1 m (3 ft.)
Termination	2 required
Cable	Shielded, twisted pair with a drain
Connector	3-pin unshielded

layer, to provide a uniform set of recommendations for the physical characteristics of the medium and the connectors. Table 9.1 shows a summary of the J1939 physical layer requirements.

It is easy to notice how the protocol requirements are considerably conservative. The required bus speed is 250 kb/s (only this bit rate is allowed). At the nominal speed, the maximum network length is 40 m (roughly 120 ft). The maximum number of nodes is 30. Compared with the typical correspondences between bus lengths and bit rates (the maximum network length at 250 kb/s, according to ISO 11898, is 250 m), these numbers are quite restrictive, probably to keep everything on the safe side and improve the robustness of J1939 systems.

9.1.2 Parameters and Parameter Groups

Standardization of message content is probably the bulk of the J1939 specification. J1939-71 defines groups of signals with related information content as *Parameters Groups* (PGs) and standardizes their definition and mapping into CAN messages. The definition covers all aspects of signal encoding: transmission rates, physical units, data types, ranges, resolutions, bit encoding and bit position. Each parameter group corresponds to a message content and largely defines the CAN identifier of the corresponding message. As an example, Table 9.2 show the definition of an engine temperature parameter group.

Standardization of message content allows reuse of subsystems that need to use or produce the same type of information data. The size of a parameter group is not limited by the maximum CAN message size of 8 bytes. If a parameter group fits within the limit, then it is mapped into a single CAN message, otherwise, the J1939 transport protocol can be used.

Out of the 1,600 pages of the SAE Truck and Bus Control and Communications Network Standards Manual (2007 Edition), approximately 1,000 pages refer to Parameter Group Assignments, Parameter Group Numbers, and Address and Identifier Assignments.

Table 9.2 The definition of engine temperature parameter group

Engine temperature PG	
Transmission rate	1 s
Data length	8 bytes
Data page	0
PDU format	254
PDU specific	238
Default priority	6
PG Number	65,262 (FEEE16)

Byte map

1	2	3	4	5	6	7	8
Engine coolant temperature	Fuel temperature	Engine oil temperature		Turbo oil temperature		Engine intercooler temperature	Not defined

Parameter groups are classified according to the transmission mode. They can be transmitted point-to-point or broadcast. Each group has a 16-bit number (PGN - Parameter Group Number) as an identifier. The PGN defines the type of the PDU (Protocol Data Unit) and is encoded in the CAN identifier of the transmitted message. The PDU and PGN format depends upon the type of the transmitted group.

If the parameter group is of type *broadcast*, then the first 8 bits of the PGN must be greater than or equal to 240 (the first four bits must be 1, that is, the hex coding must be greater than `0x0F0`). The other 12 bits of the PGN can be freely defined (leaving $2^{12} = 4{,}096$ broadcast identifiers available).

If the group is of type *point-to-point*, that is, the values are meant to be sent to specific devices, then the 8 lower order bits identify the destination address and only the first 8 bits can be used to specify the PGN (with the restriction that the value must be lower than 240). Special destination addresses are 254, indicating no node (mainly used in the address claiming procedure—see the following section on network management), and 255, indicating the entire network, which makes it a de-facto broadcast message disguised as point-to-point.

The total number of available PGNs is therefore, $4{,}096 + 240 = 4{,}336$. However, the current set of definitions are given for the Data Page bit (DP) at 0. Future extensions could be developed by providing the same number of definitions for $\mathrm{DP} = 1$, giving a total number of $4{,}336 \times 2 = 8{,}672$ PGNs (still a terribly inefficient use of 29 bits, but efficiency is clearly not one of the goals of the standard).

The current range of Parameter Group Numbers as defined in SAE J1939/71 is from PGN 0 (Torque/Speed Control) to PGN 65279 (Water in Fuel Indicator). This range is not a real representation of the total number of PGNs, since there are gaps between PGN definitions.

Of course, usage of parameter groups is not limited by the standardization in J1939. Each organization/manufacturer can use extensions to define its own manufacturer-specific parameter groups. Proprietary communication may be of type point-to-point or broadcast.

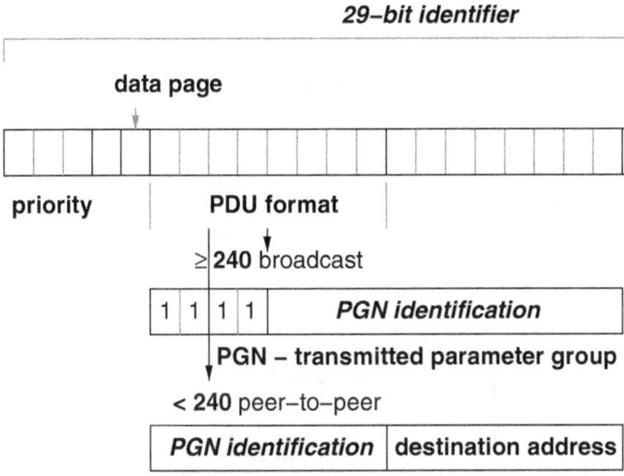

Fig. 9.2 J1939 PDU definition inside the CAN identifier

9.1.3 Protocol Data Units (PDUs)

A Protocol Data Unit is a CAN message transporting a Parameter Group as its data content and identified by a PGN. Figure 9.2 provides a graphical description of the PGN and PDU encoding into the 29-bit CAN identifier message. According to the PGN classification, a PDU can be of two types: PDU1 for point-to-point PGNs and PDU2 for broadcast. The CAN message identifier starts with three bits indicating the message priority (0-highest, 7-lowest), followed by one reserved bit and one Data Page bit. After that, the PGN (and PDU format) is defined, followed by the address (8 bit) of the source node.

In J1939, the exchange of information can be driven by a data request from the receiver, or it may happen because the sender decides autonomously and independently to initiate a communication.

- If a message is a destination-specific request or command, the receiving device filters messages for which the destination address matches one of its addresses. After processing the message, the node typically provides some type of acknowledgment.
- If a message is a global request, every device, even the originator, must process the request and respond if the data is available.
- If a message is of broadcast type, each device receives the data and then it may decide whether to use the transmitted information.

The data part of the message contains the parameter value. When a device does not have data available for a given parameter, the bytes that are allocated to the missing parameter are set to the predefined value `0x0FF` (meaning not available) so that receivers know that the sender did not have the corresponding piece of information, as opposed to a possible protocol error.

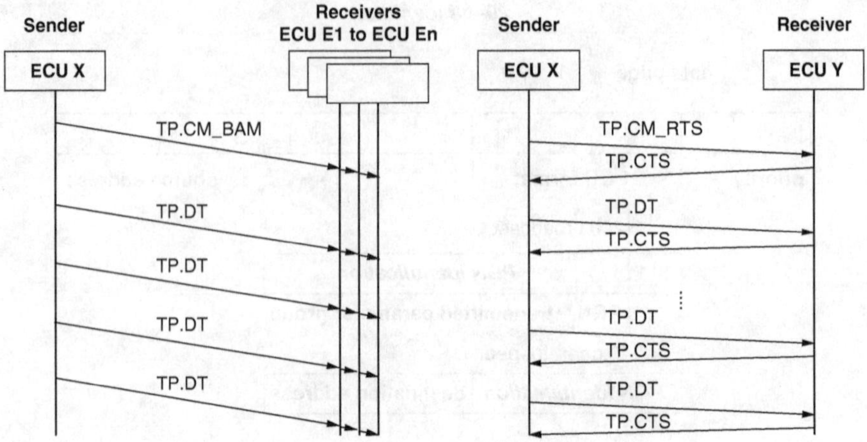

Fig. 9.3 Message sequence for a multi-packet message that is of type broadcast (*left side*) as well as point-to-point (*right side*) in J1939

9.1.4 Transport Protocol

The J1939 Transport Protocol (TP) allows to transmit large data parameter groups of up to 1,785 bytes. Messages larger than 8 bytes are fragmented at the server and packaged into a sequence of 8 byte data packets for transmission in several CAN messages. At the destination, the receiver node must re-assemble the data.

Transport protocol functions provide the necessary support for message packaging and reassembly, and for connection management. In addition, the TP functions handle flow control and handshaking for point-to-point transmissions. The control and data messages used by these functions are defined with their PGNs by the standard. In particular, Connection Management commands are assigned the PGN `0x00EC00` (60416 decimal). The specific function that needs to be performed is encoded in the Control segment of the packet. Within the transport protocol, the Data Transfer functions are performed using messages with PGN `0x00EB00` (60160 decimal).

Multi-packet transmission may be of broadcast type or point-to-point. However, multi-packet transmissions are only possible in PDU1 format. To broadcast a message longer than 8 bytes, the PDU1 format with a global destination (address = 255) must be used.

Figure 9.3 shows message sequences for a multi-packet message that is of type broadcast (left-side) or point-to-point (right side).

The sequence of messages that are required for the *broadcast transmission* of a multi-packet message is the following. The transmitter node sends a Broadcast Announce Message (BAM). A BAM message is a special type of Connection Management (CM) message (the general format of a CM PGN is in Table 9.3) characterized by a value of the Control field equal to 32. The CM_BAM message

Table 9.3 Parameters required for the definition of the connection management PGN and message format

Transport Protocol Connection Management (TP.CM)	
Parameter group number 60416 (00EC00hex)	
Definition for communication management flow-control (e.g. BAM or RTS).	
Transmission rate (according to the PGN to be transferred)	
Data length 8 bytes	
Extended data page (R) 0	
Data page 0	
PDU format 236	
PDU specific destination address (= 255 for broadcast)	
Default priority 7	
Byte index	**Data description (BAM only)**
1	Control byte (32 for BAM)
2,3	Message size (Number of bytes)
4	Total number of packets
5	Reserved `0xFF`
6-8	PGN of the multi-packet message (LSB first)

contains (among others, see Table 9.3) the description of the Parameter Group Number of the following data transmission (multi-packet message) and its size, expressed both as the number of bytes (in the Message Size field) and the number of packets. The BAM message is received by all nodes and allows those interested in the message reception to prepare for it by allocating the appropriate amount of resources (which typically consist of buffer memory). Following, the actual data content is delivered using a sequence of Data Transfer (DT) messages. The transport of Multi-Packet Broadcast messages is not regulated by any flow-control functions and thus it is necessary to define timing requirements between the sending of a Broadcast Announce Message (BAM) and the Data Transfer PGNs. The last packet of a multi-packet PGN may require less than eight data bytes, with all the unused data bytes in the last packet set to `0x0FF`.

In case of a *point-to-point communication* between two devices, the transmission starts with a CM message with a Control byte indicating Request To Send (RTS). The receiving node responds with a CM message with the Control byte indicating Clear To Send (CTS). The transmitting device then sends the portion of the data indicated in the CTS using DT messages. This handshake of CTS messages with a response consisting of DT messages continues until the entire set of required data (higher-level protocol data unit) is transmitted. The connection is terminated at the completion of the message by the receiver transmitting a CM message with a Control byte indicating End Of Message Acknowledgement (EOM). Note that for this process to work, the CM message contains additional data based on what the control byte is. The RTS includes: number of bytes, number of packets, and the PGN data that is carried in the data part of the communication. The CTS includes the number of packets the receiver expects next and the next packet number.

Table 9.4 Definition of an ECU name in J1939

Definition of the 64 bits in the ECU name		
Byte number in CAN message	Contents/Meaning	Field size
0	Identity number, LSB	21
1	Identity number	
2 Bits 0-4	Identity number, MSB	
2 Bits 5-7	Manufacturer code, LSB	11
3	Manufacturer code, MSB	
4 Bits 0-2	ECU instance	3
4 Bits 3-7	Function instance	5
5	Function	8
6 Bit 0	Reserved bit	1
6 Bits 1-7	Vehicle system	7
7 Bits 0-3	Vehicle system instance	4
7 Bits 4-6	Industry group	3
7 Bit 7	Arbitrary address bit	1

9.1.5 Network Management

The most important function in J1939 Network Management is the Address Claiming Process. Each ECU in a J1939 vehicle network is uniquely identified by a 64-bit *name* that also indicates the ECU main function. The ECU name is indeed a structure, and it is composed of ten fields, as shown in Table 9.4. An ECU name is statically assigned by the manufacturer to each J1939 device.

The ECU name is not meant to be used as an address inside the CAN messages (it would be clearly impractical since it would consume all the 8 bytes in the CAN data field). For the purpose of expressing an address inside a PDU (a message), each ECU also has a shorter 8-bit address. ECU addresses are typically predefined, assigned statically and used immediately upon power up. However, the SAE J1939 standard also supports the dynamic assignment of addresses to nodes, to accommodate future devices and functions which have not yet been defined. In this case, an address claim procedure is defined to be executed right after the network initialization is completed. The procedure assigns addresses to ECUs and, of course, it ensures that they are unique throughout the network. J1939 states that support for the dynamic address assignment procedure is optional and only those devices which might be expected to encounter address conflicts must support this capability. To eliminate the need to support dynamic address assignment and speed up this identity process, most ECUs are associated with a preferred address, as described in the standard document J1939/71.

Two options are available to an ECU for claiming an address. In the first case, the ECU first sends a Request for Address Claim message, which allows it to learn all the addresses of the other nodes in the network before claiming its own.

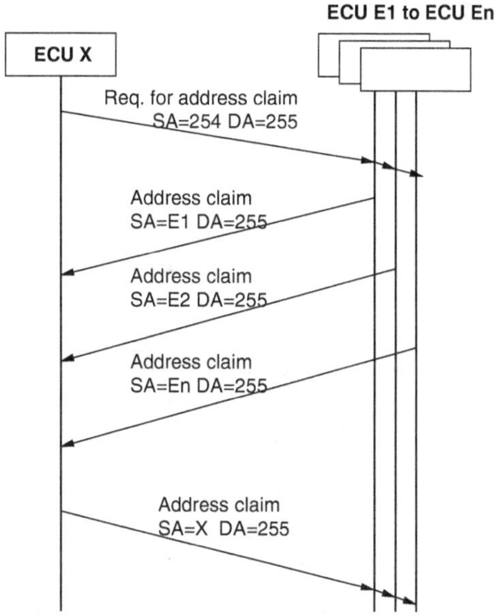

Fig. 9.4 J1939 address claiming procedure starting with a Request for Address Claim

Alternatively, the ECU may immediately try to claim one 8-bit address and deal with the possible address collision.

- *Request for Address Claim*. In this case, the device that needs an address sends a Request for Address Claim, that is, a special broadcast message (with destination address equal to 255, see Fig. 9.4). The requesting node does not have an address yet and uses the NULL address (254) as its source address in the PDU. All the other devices in the network respond by transmitting their addresses inside an Address Claimed J1939 message. This allows the requesting device (typically a transitional devices such as a tool or a trailer or a device that powered up late) to obtain the current table of all the addresses in use in the network so that it can select a free address and claim it using an Address Claim Message (see Fig. 9.4).
- *Address Claim Message*. The destination address for an address claim is the global address (255) to broadcast to all nodes. The requesting node is now using the claimed address as its source address and sends its name as the data content of the message. All receiving devices compare this newly claimed address to their own table of devices in the network. In the event that more than one ECU attempts to claim the same address, the ECU with the lowest name has the highest priority. In case one of the receivers has a conflicting address and higher priority, it notifies the sender with an Address Claim Message indicating that the address is already in use. The remaining ECUs must claim a different address by sending another Address Claimed Message containing a different address or send a Cannot Claim Address Message.

9.2 CANopen

The CANopen standard includes a set of protocols and recommendations spanning several layers, from the physical layer to the application level. In several ways, it is similar to J1939, since it includes not only the definition of connectors, cables and bit time specifications at the physical level, but also higher level protocols, including a Transport-level protocol, Network Management functionality, and standard services, such as time and synchronization services, and especially a set of standard definitions for application-level objects grouped into a set of application profiles and device-specific profiles.

CANopen was initially developed in the context of a European project, with Bosch acting as the coordinator. In 1995, the CANopen specification was taken over by the CAN in Automation (CiA) group and then standardized as EN 50325-4. The CANopen specifications are freely available to the public after registration on the CiA web site http://www.can-cia.org/. An additional excellent source of information is available on the web at http://www.softing.com/, and an open source implementation of CANopen can also be found at http://www.canfestival.org/ (with quite a few limitations). Readers interested in the details of CANopen and possibly their practical implementation can find additional information in [25] and [52].

9.2.1 CANopen Architecture and Standards

The overall structure of the CANopen specifications can be represented as in Fig. 9.5. The standards for the lower layers ensure that each CANopen device adopts the same communication interface and protocol software. Lower layers provide the communication infrastructure for the high level services and the transmission and reception of application-level communication objects over the bus. Objects are defined in an *object dictionary*, organized by device. The dictionary also describes all the data types and provides the CANopen interface to the application software.

In the following sections, we discuss in more details the standard requirements at each level, starting from the Physical layer.

9.2.2 Physical Level

At the physical level, CANopen complements the CAN specifications with a set of definitions of acceptable wiring connections, bit time specifications and connector specifications. The physical medium for CANopen devices is a differential two-wire bus line with common return according to ISO 11898.

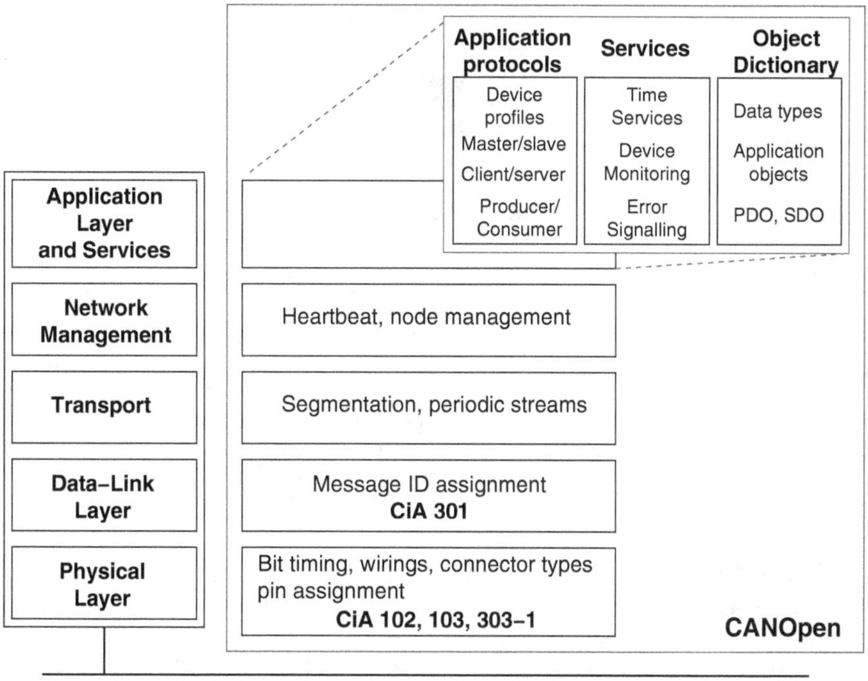

Fig. 9.5 The CANopen standards, from the application to the physical layer

Table 9.5 Acceptable bit rates, bus lengths, and bit timings as specified by CANopen

Bit rate	Bus length	Bit time	Time quantum	Num of quanta	Bit sample time
1Mb/s	25 m	1 it μs	125 ns	8	6
800 kb/s	50 m	1.25 it μs	125 ns	10	8
500 kb/s	100 m	2 it μs	125 ns	16	14
250 kb/s	250 m	4 it μs	250 ns	16	14
125 kb/s	500 m	8 it μs	500 ns	16	14
50 kb/s	1,000 m	20 it μs	1.25 it μs	16	14
20 kb/s	2,500 m	50 it μs	3.125 it μs	16	14
10 kb/s	5,000 m	100 it μs	6.25 it μs	16	14

9.2.2.1 Bit Rates and Bit Sampling Times

CANopen has a set of possible bus transmission rates from 1Mb/s to 10kb/s, with the corresponding maximum bus lengths and a set of definitions for the time quantum size and the bit sampling instant associated to each rate. At 1 Mbit/s, each bit time consists of 8 time quanta, at 800 kbit/s of 10 time quanta, and from 500 kbit/s to 10 kbit/s of 16 time quanta. The bit sampling times (CANopen uses the single sampling mode) are defined accordingly at 6/8 of the bit time, 8/10 and, from 500 kb/s down, at 7/8 of the bit time. These values are shown in Table 9.5.

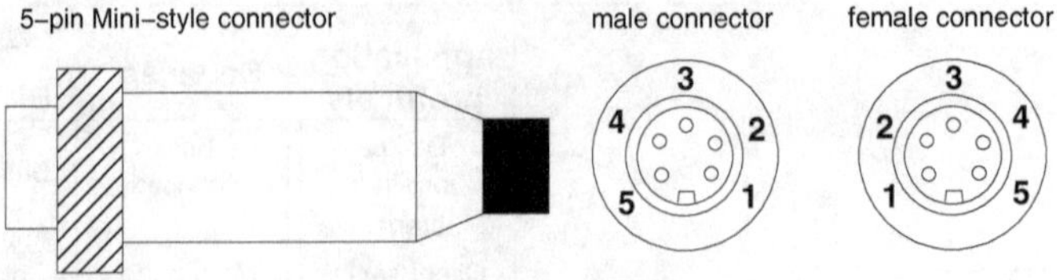

Fig. 9.6 The CANopen pin connections for a 5-pin mini connector

Table 9.6 Pin assignment for the 5-pin mini connector

Pin	Signal	Description
1	CAN_SHLD	(Optional) CAN shield
2	CAN_V+	(Optional) CAN external positive supply voltage for transceivers and optocouplers
3	CAN_GND	Ground / 0V
4	CAN_H	High bus line
5	CAN_L	Low bus line

Every CANopen module has to support at least one of the specified bit rates and possibly more than one. Maximum bus lengths are computed on the basis of an estimated worst-case 5 ns/m propagation delay and a total device delay (on both the transmission and reception side) of 210 ns for rates between 1Mb/s and 800 kb/s, 300 ns (including the delay on the recommended optocouplers for bus lengths longer than 200 m) between 500 and 200 kb/s, 450 ns at 125 kb/s, and 1.5 time quanta at lower rates. For bus lengths greater than 200 m the use of optocouplers is recommended. If the bus is longer than 1,000 m, bridge or repeater devices may be needed.

9.2.2.2 Connectors

The CiA DRP-303-1 CANopen recommendation defines the pin assignments for the 9-pin D-sub connector (DIN 41652 or corresponding international standard), the 5-pin mini style connector, the open style connector, the multi-pole connector, and other connectors. The 9-pin D-Sub connector pin assignment complies to a CiA standard denoted as DS-102. The association of the CAN signals with the connector pins has been shown as an example of additional physical level specification in Chap. 1. As a further example, Fig. 9.6 and Table 9.6 show the pin connections for a 5-pin mini-style connector.

9.2.3 Object Dictionary

Each CANopen device maintains an object dictionary containing the description of communication and application objects as well as data type definitions. Each

Index	Object type
0000	Reserved
0001–009F	Data types
00A0–0FFF	Reserved
1000–1FFF	Communication Profile
2000–5FFF	Manufacturer Specific Profile
6000–9FFF	Standardized Device Profile
A000–FFFF	Reserved

Fig. 9.7 The CANopen Object table index

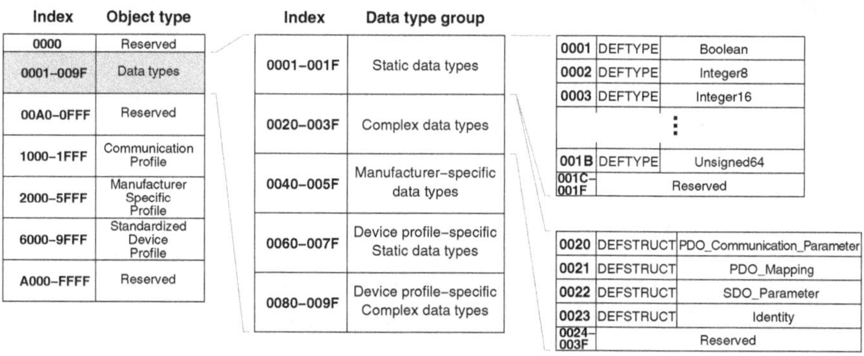

Fig. 9.8 The CANopen Object table indexes for object types

dictionary item is addressed using a 16-bit index and an optional 8-bit sub-index. The pairing of each object to its index is partly defined by the standard and partly available for manufacturer-specific extensions. Also, indexes can be assigned statically or as the result of a configuration procedure at system startup. The overall mapping of the object dictionary items with the corresponding first-level 16-bit indexes is shown in Fig. 9.7.

9.2.3.1 Type Entries

The first entries of the data dictionary are reserved to a set of common definitions of static data types, as shown in Fig. 9.8. Definitions include common types, such as

Index	Object type
0000	Reserved
0001–009F	Data types
00A0–0FFF	Reserved
1000–1FFF	Communication Profile
2000–5FFF	Manufacturer Specific Profile
6000–9FFF	Standardized Device Profile
A000–FFFF	Reserved

Index	Object	Name	Type	Acc	M/O
1000	VAR	Device type	Unsigned32	RO	M
1001	VAR	Error register	Unsigned8	RO	M
1002	VAR	Manufacturer status register	Unsigned32	RO	O
1003	ARRAY	Predefined error field	Unsigned32	RO	O
1004		Reserved			
1005	VAR	COB–ID, SYNC–message	Unsigned32	RW	O
⋮					
1018	RECORD	Identity object	Identity	RO	M
⋮					
1FFF		Reserved			

Fig. 9.9 The CANopen Object table indexes for the Communication Profile Objects

Bit index	7	6	5	4	3	2	1	0
	Generic error	Current	Voltage	Temperature	Communication error	Device–profile specific	Reserved (0)	Manufacturer–specific
Mandatory/Optional	M	O	O	O	O	O	O	O

Fig. 9.10 The error object register bits and their meaning

Boolean, IntegerX (where X is the number of bits, from 8 to 64), UnsignedX, Float, Visible string, Octet string, Date, Time of day, and Time differences. CANopen specifies some predefined complex data types for communication and protocol parameters (called PDO and SDO parameters). In addition, the object dictionary reserves entries for device-specific standard and complex data types.

9.2.3.2 Object Descriptions

Each Object is defined in CANopen by a set of attributes, including its *Name*, which also provides a description of the object and its intended use; an *Object Code*, a *Data Type*, and its *Category*, which specifies whether its implementation is mandatory, optional, or conditional (Fig. 9.9). The Object Code must be one of those defined by the CANopen specification, such as Variable, Array, or Record. For simple variables, the description is immediately available in the dictionary at the matching index entry. For complex data types (such as Arrays and Records), the index refers to the entire data structure, and each attribute is individually described, and identified by the sub-index field.

The object at index 1001h is an *Error Register* for the device (Fig. 9.10). This object entry is mandatory for all devices as part of the Emergency object. The error object is an 8-bit register, which records the occurrence of internal errors

Index	Subindex	Description	Data type
1018	0	Number of entries	Unsigned8
	1	Vendor ID	Unsigned32
	2	Product code	Unsigned32
	3	Revision number	Unsigned32
	4	Serial number	Unsigned32

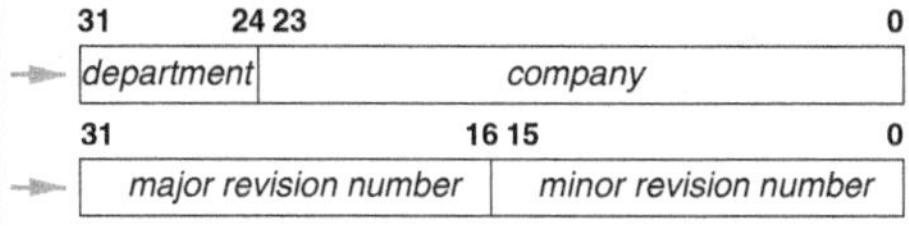

Fig. 9.11 The CANopen Identity Object structure with the bit map for the fields Vendor ID and Revision number

(at the application level). If a bit is set to 1, then the corresponding error has been detected. The only mandatory error bit is the generic error, which must be signaled in any error situation.

Another mandatory object is the Identity Object at index 1018h (Fig. 9.11). This is an example of a structured object, which contains general information about the device. The *Vendor ID* is a numeric value of type Unsigned32 and consists of a unique number for each registered company and possibly department within the company (if required). Vendor IDs are assigned and kept in a worldwide register managed by the CiA. Registration comes at a small fee (waived for CiA members). The manufacturer-specific *Product code* identifies a specific device version. The *Revision number* consists of a major and a minor revision number, both identifying a given device behavior. If the device functionality is updated, the major revision number must be increased.

9.2.4 Time Services

CANopen provides several services related to time. It provides a time synchronization service based on a special set of objects and messages. This service uses the object protocol (discussed in the next sections), but also provides the foundation for the definition of synchronous and cyclic messages. This is why it is discussed early in this chapter. CANopen also defines a timestamping service to provide network nodes with a system-wide global clock.

9.2.4.1 Synchronization Service

In the synchronization service, one node acts as the synchronization master and is the producer of the Synchronization Object. The other devices are slaves or consumers. The purpose of the Synchronization Object is not to build a global clock, but to allow network devices to synchronize their actions. When slave devices receive the synchronization message, they start carrying out their synchronous activities (for example, sensor sampling or actuations) or message transmissions in a coordinated fashion.

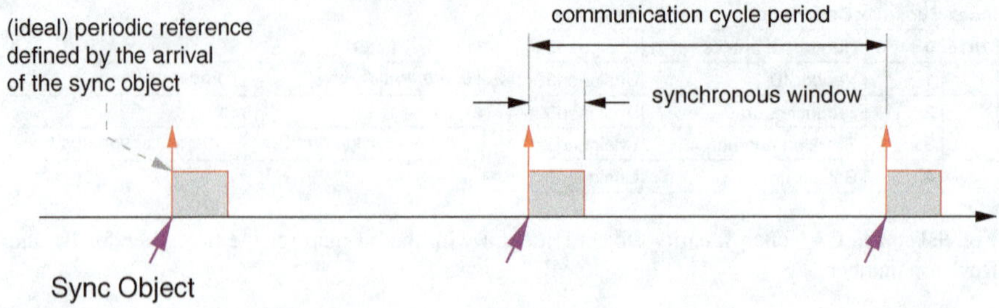

Fig. 9.12 Synchronization message, with the corresponding period and synchronization window

The identifier of the Synchronization Object is available in the object dictionary at index 1005h. The Synchronization Object does not carry any data. In order to provide maximum possible accuracy in the periodic time reference it provides, interference from other messages should be avoided and the synchronization message should have a very high priority CAN identifier. CANopen suggests the identifier 128 which is a value of the highest priority group used in the CAN identifier assignment (see following sections). Still, there can be time jitter in the transmission of the message because of the possible blocking time caused by a lower priority message being transmitted just before the Synchronization Object. Also, in case of a transmission error, the synchronization message may be retransmitted and received with a slight delay.

Synchronization of node activities (including the transmission of object messages implementing periodic streams of data) with respect to the time reference provided by the Synchronous object requires additional definitions. The time period between two consecutive transmission of Synchronization Objects is specified by the parameter *communication cycle period*. Also, a time window denoted as *synchronous window length* is defined starting at the time the Synchronization message is transmitted, once for every message period (Fig. 9.12). The *synchronous window length* (at index 1007h) and the *communication cycle period* (index 1006h) are specified in the Object Dictionary, which may be written by a configuration tool to the application devices during the boot-up process.

9.2.4.2 Time Stamp

The Time Stamp object is transmitted from a given network node acting as the service master to the other network nodes (Fig. 9.13), providing an absolute time reference, expressed as the number of ms elapsed since midnight of January 1st, 1984. Since the number is expressed as a 48 bit integer (6 bytes), it will suffice for almost 9,000 years.

Similar to the synchronization object, the time stamp object also needs a high priority identifier, in order to be received with minimum jitter. CANopen suggests to use the identifier 256.

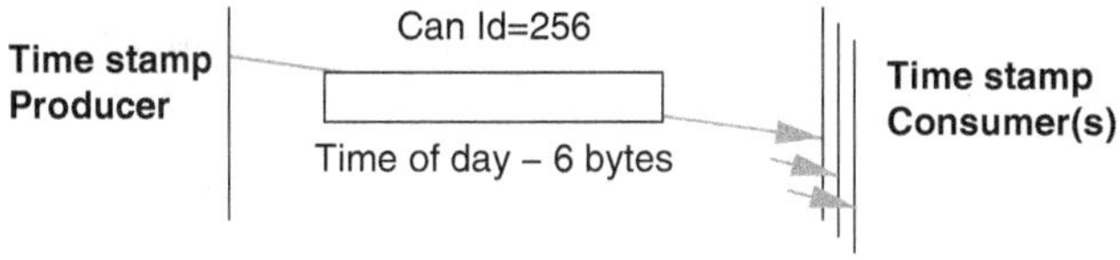

Fig. 9.13 The time stamp object

Some time critical applications may still require a tighter synchronization, below the millisecond-precision of the standard time stamp service. For this purpose, CANopen provides an optional high resolution synchronization protocol which employs a special form of time stamp message to be used together with a clock synchronization algorithm (that tries to compensate for the local clock drifts). The high-resolution time-stamp is encoded as an unsigned integer with 32 bits with a resolution of 1 microsecond, which provides about 72 min of clock lifetime or wraparound time. The high resolution time-stamp is mapped into a Protocol Data Object (see the following section on communication protocols) with object index 1013h.

9.2.5 Communication Protocols

CANopen has several protocols for the purpose of transmitting information on objects over the network. Since objects may be larger than the maximum payload of a CAN message, object protocols include several functions that are typical of the transport layer, such as the possibility of message fragmentation and reassembly, frame-level flow control and synchronization. In addition, communication protocols make use of the previously described time services and also provide for emergency management. The object protocols can be classified as follows

- Protocols for service-oriented communication. These protocols are used for read and write access to entries of a (remote) device object dictionary. They make use of Service Data Objects (SDO).
- Protocols for (real-time) data transfers. These protocols make use of Process Data Objects (PDO).
- Special Object Protocols, including Synchronization (SYNC), Time Stamp, and Emergency (EMCY) Protocols.

Another set of protocols is dedicated to the Network management and is discussed in a separate section.

9.2.6 Service-Oriented Communication and SDO

Service Data Objects (SDO) provide access to the entries of a device object dictionary. A client device initiates the transfer. Each SDO communication requires

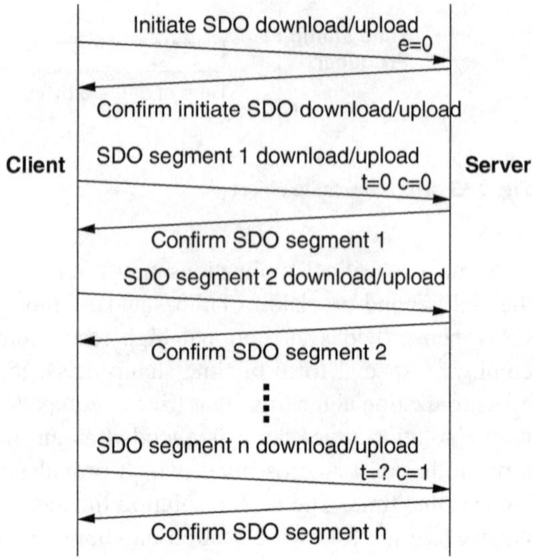

Fig. 9.14 The transfer protocol for SDO downloads and uploads

two CAN data frames with different identifiers, one for the outgoing request by the client, and the other for the required confirmation from the owner of the accessed dictionary, acting as the server. The reception of each segment is acknowledged by a corresponding CAN message.

SDOs allow to transfer data objects of any size, fragmented as a sequence of segments of no more than 8 bytes. Actually, since the SDO connection is initiated with a message that has 4 spare bytes in its payload, for messages of up to 4 bytes an *expedited transfer* may be used, performing the data transmission within the message that initiates the transfer protocol. Messages of more than 4 bytes must be transmitted by means of the segmented transfer, split up over several CAN messages (as shown in Fig. 9.14).

When reading from or writing to the object dictionary of the server device, the service Initiate SDO Download or Upload message is the first to be transmitted. Its format (with the corresponding response message by the server) is detailed in Fig. 9.15 (data bytes are transmitted with the most significant bit first). The first byte is the *Command Specifier*, followed by the *Index* and *Subindex* fields, indicating the object table entry for which the read/write operation is requested (Fig. 9.16), and a payload of up to 4 bytes.

The Command Specifier of the request message from the client contains the following information.

- *S*: one bit (at position 0), defining whether the transfer protocol is of block type or segmented type.
- *E*: one bit (position 1), defining whether the protocol is an expedited transfer.
- *N*: two bits (positions 2 and 3), indicating the number of bytes in the data field.
- *X*: unused bits.
- *CCS*: (positions 5 to 7) Client Command Specifier.

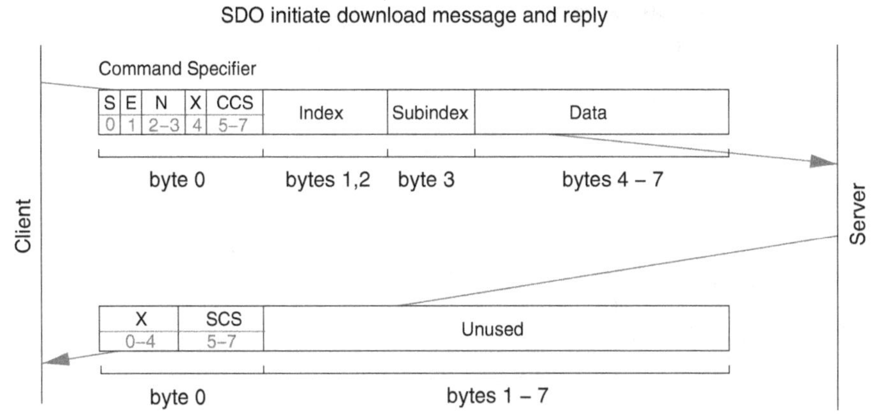

Fig. 9.15 SDO Transfer Initiate message and its reply from the server

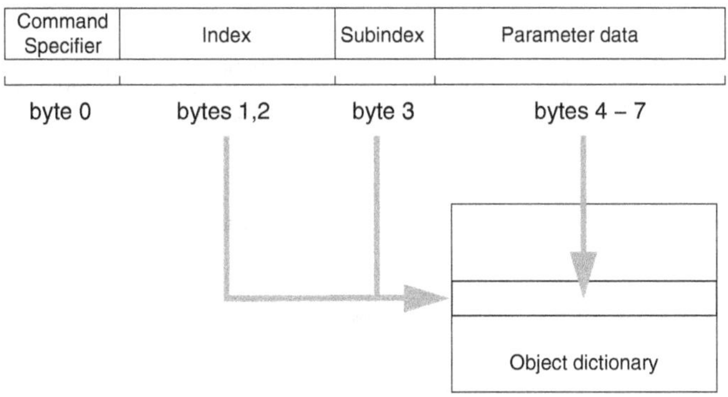

Fig. 9.16 The initiate transfer message indicates the object dictionary entry on which to operate

In the server reply, the only field of interest is the SCS, or Server Command Specifier. Also, the CAN Id of the request message (see also the following section on the assignment of the CAN identifiers) tells whether the request is for a download or upload (and of course allows to identify the response message for a given request).

All segments after the first CAN message may each contain seven bytes of data. For each SDO segment, two CAN Data Frames are exchanged between the client and the server. The request carries the segment of data, as well as segmentation control information. The response acknowledges the reception of each segment. In the Command Specifier field of the segment frame, two additional fields are defined.

- *C*: one bit in the request frame (at position 0), an indicator of the last segment (or end indicator, see Fig. 9.17).
- *t*: one bit in the reply frame (at position 4), toggled on successive segment frames starting with an initial value of zero.

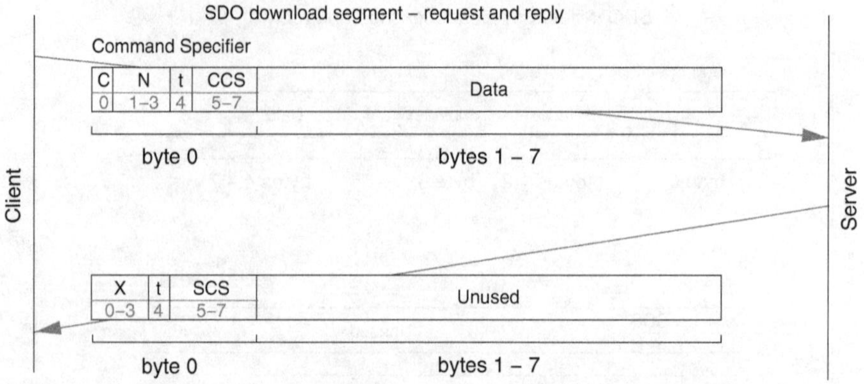

Fig. 9.17 The SDO format for segmented transfers

Fig. 9.18 SDO download block exchange

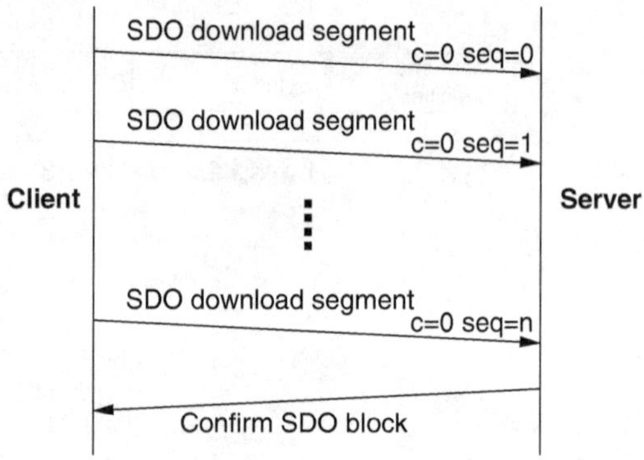

Both the client and the server can abort the transfer by sending an appropriate message.

Optionally an SDO can be transferred as a sequence of blocks without individual acknowledgements (Download SDO Block, Fig. 9.18). In this case, each block is a sequence of up to 127 segments containing a sequence number and the data. The entire message sequence is confirmed by only one reply message from the server, at the end. If in the confirmation message the *C*-bit is set to 1, the block download sequence was received successfully.

9.2.6.1 Data-Oriented Communication and PDO

The PDO communication is a data oriented communication, in which a node acts as the producer of the information and the other is the consumer. Optionally, there can be many consumers and the transmission is of broadcast type. Process Data Objects are always transmitted without acknowledgement from the receiver (they are sent as datagrams). The transmission can be initiated by the producer (in case of write-PDO) or by the consumer (in case of a read-PDO). In addition, the transmission of data can be aperiodic, sporadic, or periodic (Fig. 9.19).

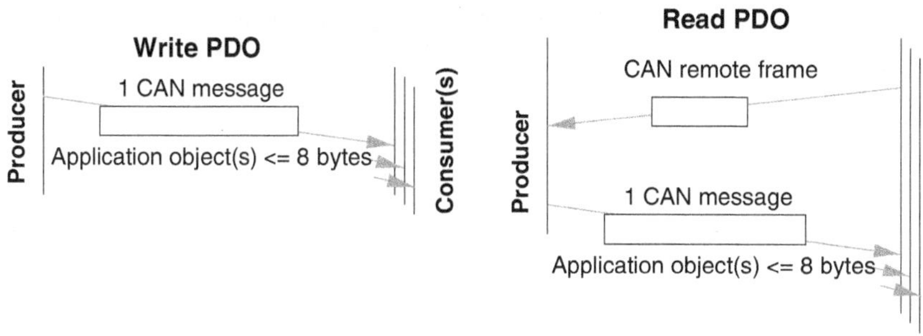

Fig. 9.19 Message exchanges for Read- or Write-type PDOs

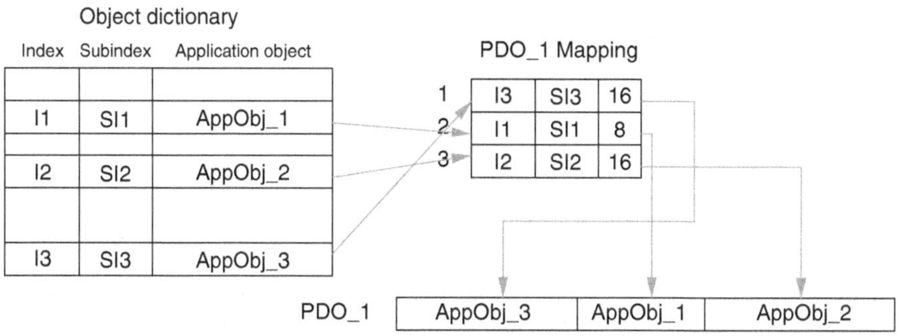

Fig. 9.20 Mapping of application objects into PDOs

Write-PDO A protocol data object for transmission in write mode (initiated by the producer) is always mapped to a single CAN Data frame and therefore cannot be longer than 8 bytes. The data field of the CAN message may contain any type of application data or application objects (see the following section on the mapping of CANopen objects to PDOs). Each device has a finite number of available PDOs. Their number and length are application-specific and have to be specified in the device profile. In one CANopen network there can be up to 512 Write- or Transmit-type PDOs (T_PDOs) and 512 Read PDOs (R_PDOs).

The PDOs correspond to entries in the Object Dictionary and provide the interface to application objects. The data type and the mapping of application objects into a PDO is determined by the PDO mapping structure within the Object Dictionary.

Read-PDO The Read-PDO is mapped to a CAN Remote Frame, which is responded by the corresponding CAN Data Frame. The implementation of Read-PDOs is optional.

Object-to-PDO Mapping Inside each device, a mapping can be defined to link the data content of each PDO to the application objects in the dictionary. The mapping (a representation is shown in Fig. 9.20) defines the index (and subindex) of the mapped objects, their position in the PDO data field, and their size.

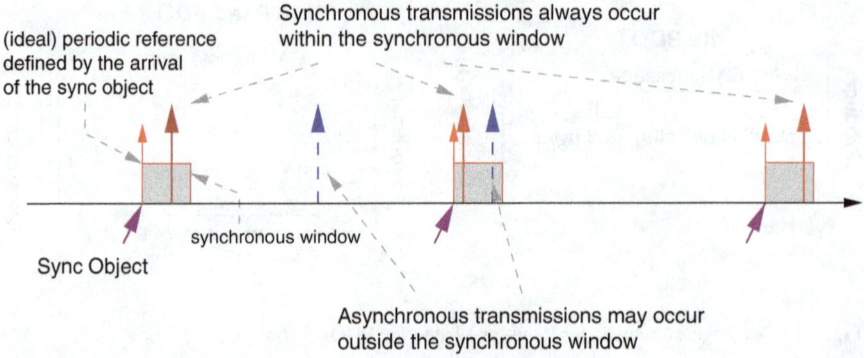

Fig. 9.21 Synchronous and asynchronous PDO messages

Fig. 9.22 A taxonomy of PDO transmission modes

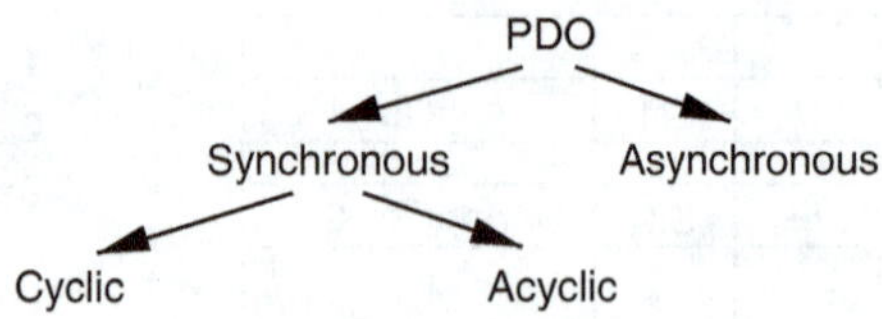

Trigger Modes The CANopen communication profile defines three message triggering modes for PDOs. The first two are of type Asynchronous, the third is of type Synchronous.

- The transmission of a PDO message is triggered by the occurrence of an object-specific event defined in the device profile. The transmission can be optionally triggered by a device timer even if no event occurs (realizing periodic streams without network synchronization).
- The transmission of asynchronous PDOs may be initiated on the reception of a remote request initiated by another device.
- Synchronous PDOs are triggered by the expiration of a specific transmission period synchronized by the reception of the SYNC object.

Synchronous PDOs are transmitted within the synchronous window after the Synchronization Object (Fig. 9.21). The priority of synchronous PDOs should be higher than the priority of asynchronous PDOs to ensure minimum jitter. Asynchronous PDOs and SDOs can be transmitted at any time (outside or inside the synchronous window). Synchronous PDOs are further classified according to the transmission mode as cyclic or acyclic (Fig. 9.22).

Synchronous cyclic PDOs are periodic messages transmitted within the synchronous window. The number of the transmission type (1 to 240) indicates the message period, expressed as the number of Synchronization objects transmitted between two cyclic PDOs. Acyclic synchronous PDOs (of type 0) are triggered by an application event. The message must be transmitted within the synchronization window but is not necessarily periodic (Fig. 9.23).

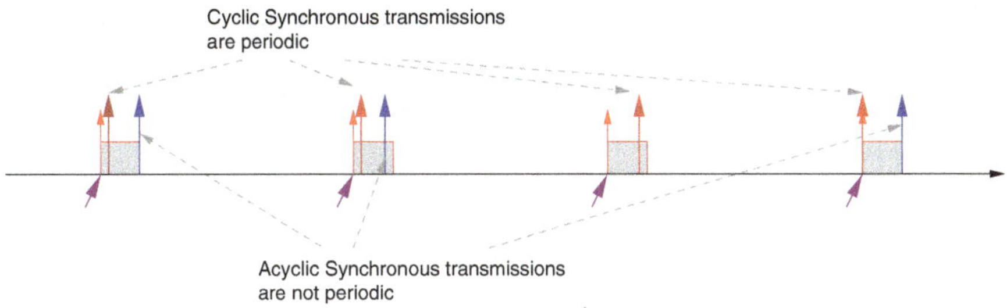

Fig. 9.23 Cyclic and acyclic synchronous PDOs

Sporadic Transmissions with a Minimum Intertransmission Time To ensure that a stream of asynchronous PDOs does not jeopardize the transmission of other lower priority messages, CANopen allows to enforce a minimum intertransmission time (or transmission inhibit time) for PDOs, that is, the minimum time that must elapse between any two consecutive transmissions of PDO service requests.

9.2.7 Message Identifier Allocation

CANopen defines a mandatory scheme for the allocation of message identifiers. Each message identifier is partitioned into two subfields. The first 4 bits of the identifier define the message function (Function Code). The remaining 7 bits define the Module-ID part, which identifies the source device. The Module-ID may be assigned in several ways, either statically (for example, using DIP switches) or using a dedicated service (which is part of the Network Management). Table 9.7 shows how the identifiers of the CAN messages used by CANopen are constructed based on the function that the message is performing and the identifier of the sender or receiver device (Fig. 9.24).

9.2.7.1 Emergency Object

Emergency messages are triggered by the occurrence of a critical error for a device. They are transmitted from the device detecting its application error to all other devices, typically with high priority. Emergency objects must be sent only once for each detected error event. The Emergency Object contains 8 data bytes and is acknowledged by the receiving devices. The Emergency Object is optional, but if a device supports it, then it must support at least the two error codes 00xx (error reset or no error) and 11xx (generic error).

Table 9.7 Definition of the CAN identifier based on the function code and the device identifier

Function code (Bin)	Device (Hex)	CAN Id	Communication objects
0000	0	0	NMT service
0001	0	80h	SYNC message
0001	1h - 7Fh	81h - FFh	Emergency messages
0010	0	0	Time stamp message
0011	1h - 7Fh	181h - 1FFh	1st Transmit PDO
0100	1h - 7Fh	201h - 27Fh	1st Receive PDO
0101	1h - 7Fh	281h - 2FFh	2nd Transmit PDO
0110	1h - 7Fh	301h - 37Fh	2nd Receive PDO
0111	1h - 7Fh	381h - 3FFh	3rd Transmit PDO
1000	1h - 7Fh	401h - 47Fh	3rd Receive PDO
1001	1h - 7Fh	481h - 4FFh	4th Transmit PDO
1010	1h - 7Fh	501h - 57Fh	4th Receive PDO
1011	1h - 7Fh	581h - 5FFh	Transmit SDO
1100	1h - 7Fh	601h - 67Fh	Receive SDO
1110	1h - 7Fh	701h - 77Fh	NMT error control

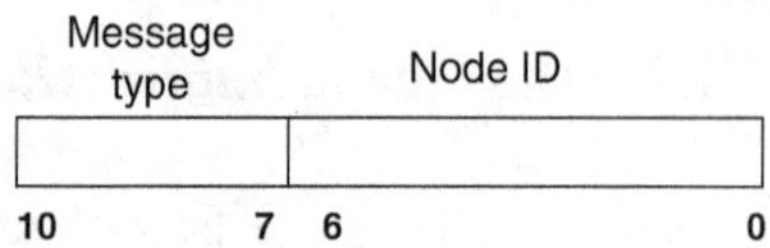

Fig. 9.24 The CANopen Object table indexes for object types

9.2.8 *Network Management*

The Network ManagemenT (NMT) protocols provide services for network initialization, error control and device status control. Overall, the network management functions include a true NMT protocol, a Boot-Up protocol, and an error control protocol.

The CANopen network management is node-oriented and follows a master/slave structure. One device in the network has the role of NMT master. The others are slaves, uniquely identified by their device identifiers. The network management provides the following functionality groups:

- *Module Control Services* for the initialization of all NMT slaves.
- *Error Control Services* for the supervision of the nodes and the network communication status.
- *Configuration Control Services* for uploading or downloading configuration for a network device.

All NMT slave devices are required to implement the state machine defined in Fig. 9.25, which automatically brings every device in the *Pre-operational* state after power-on and internal initialization. In this state, the node may be configured and parameterized via SDO (e.g. using a configuration tool) and no PDO communication

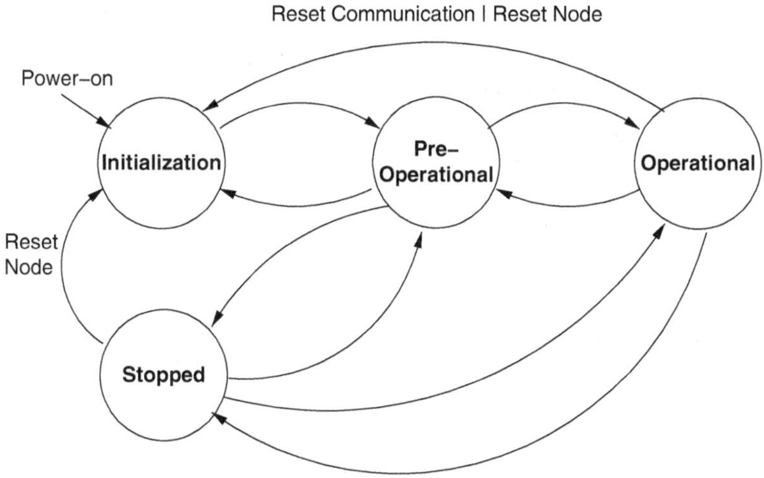

Fig. 9.25 The device states controlled by the Network Management protocol

is allowed. The NMT master may switch all nodes or a single node to the *Operational* state or bring them back to Pre-operational or even to a *Stopped* state. In the Operational state, the exchange of PDOs is allowed. Devices in Stopped state cannot communicate in any way (neither PDO, nor SDO). In the Operational state, all communication objects are active. However, the device application code may selectively allow read and write access only for some objects in the dictionary, In detail, the commands that can be issued by the NMT master are the following (identified by the Command Specifier or CS)

- Start Remote Node (CS = 1)
- Stop Remote Node (CS = 2)
- Enter Pre-Operational (CS = 128)
- Reset Node (CS = 129)
- Reset Communication (CS = 130)

The communication object has an identifier equal to 0. The Node-ID defines the destination of the message. If it is zero, then the protocol addresses all NMT slaves.

The Initialization state is divided into three sub-states (Fig. 9.26) in order to enable a complete or partial reset of a node. In the *Reset Application* sub-state the parameters of the manufacturer-specific profile area and of the standardized device profile area are set to their default values. After setting the power-on values, the *Reset Communication* sub-state is entered. In this sub-state, the parameters of the communication profile are set to their power-on values. After, the *Initialization* sub-state is entered, and the basic device initialization is executed. Before entering the Pre-Operational, the Boot-up Object is transmitted.

The Bootup protocol is used to signal that an NMT slave has entered the node state Pre-Operational after the state Initializing. The protocol uses the same identifier as the error control protocols. The boot-up message is transmitted also after reset-communication, reset-application and recovering from power-off.

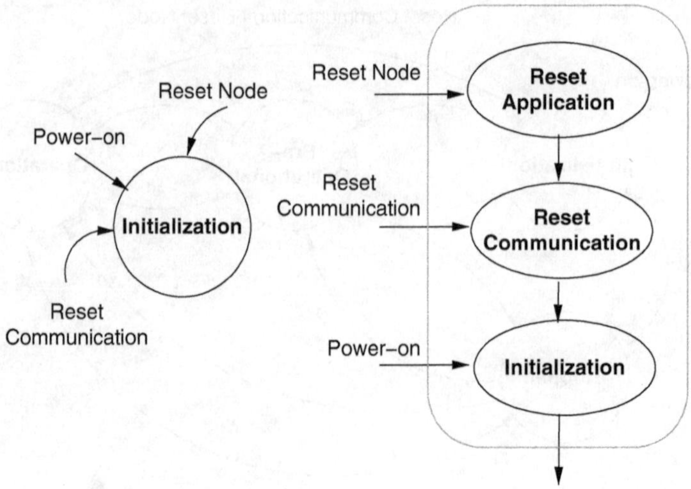

Fig. 9.26 The three sub-states in initialization state

9.3 CCP

The CAN Calibration Protocol (CCP) [8] is intended especially for calibration and recording measurement data in control units. The CCP has been defined by the European ASAP Task Force (now renamed to ASAM) as a high-speed interface based on CAN for measurement, calibration, and diagnostics systems. There is no requirement on the choice of the Physical Layer or the communication bit rate. The current version 2.1 of the CCP calibration protocol provides the following features for the development and testing phase as well as end-of line programming:

- Read and write access to a very wide variety of memory locations.
- Simultaneous calibration and measurement.
- Flash programming.
- Simultaneous handling of different control units.

9.3.1 Protocol Definition

The CCP is a master-slave type communication. The master device, typically the tool for calibration/diagnositics/measurement purposes hosted on a PC, initiates the data transfer on the CAN bus by sending commands to the slave devices, the ECUs connected to the CAN bus. The tool sends initialization messages informing which measurement data needs to be sent to the tool, as well as information on whether the measurement data is event-driven or periodically sampled. Once this message has been received, the slave responses with a single CAN message. The ECU automatically sends the measurement data at the appropriate time after initialization.

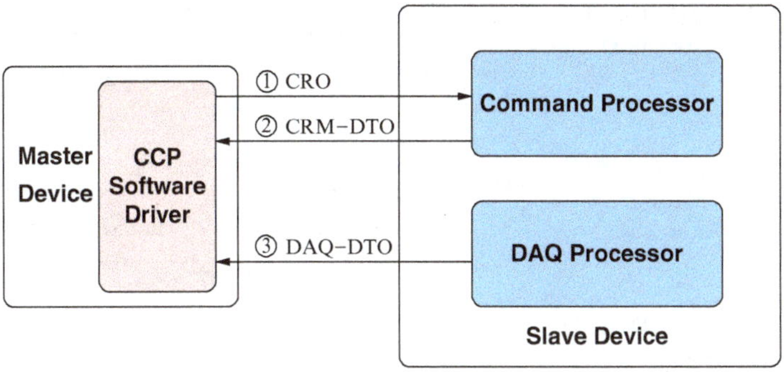

Fig. 9.27 Communication flow between CCP master and slave devices

The master sends a command contained in a single CAN message, which is defined from the slave's point of view as the Command Receive Object (CRO). The message sent from the slave to the master is also defined from the slave's point of view as the Data Transmission Object (DTO). The message sent by the slave in direct response to a CRO is called the Command Return Message (CRM), which can be a simple acknowledgement to the CRO, or can contain the actual requested data. The other two types of DTO are the Event Message (a specific type used to inform the master of the internal status changes caused by events), and the Data Acquisition (DAQ) message (used to send measurement data to the master). The typical communication flow between the tool (the master) and the ECUs (the slaves) is illustrated in Fig. 9.27. On the CCP master, the CCP Software driver functions as an interface between the application and the CAN Driver. The CAN Driver receives the CAN messages from the devices and forwards the data information contained in the CCP CAN message to the CCP Software Driver.

The master device is configured with the slave device description file, which defines the assignment of message identifiers to message objects. The identifiers (priorities) of the calibration messages should be defined in a way such that the real-time constraints of all the messages on the bus are satisfied.

9.3.2 *Command Receive Object*

The Command Receive Object (CRO) is used for transmission of command codes and related parameters from the tool to the ECU(s). The slave device answers with a Data Transmission Object containing a Command Return Message. The data length of CRO must always be 8 bytes. As shown in Fig. 9.28, byte 0 of a CRO is the command code (CMD), byte 1 is the command counter (CTR), and bytes 2 to 7 contain the command-related parameter and data.

Fig. 9.28 Message format of a Command Receive Object

Byte	0	1	2...7
	CMD	CTR	Parameter and data field

Fig. 9.29 Format of a Command Return Message or an Event Message

Byte	0	1	2	3...7
	PID	ERR	CTR	Parameter and data field

Fig. 9.30 Format of a Data Acquisition Message

Byte	0	1...7
	PID	Parameter and data field

9.3.3 Data Transmission Object

The Data Transmission Object (DTO) is used to send information from the ECU to the tool. Three types of DTOs are defined by the CCP specification: the Command Return Message (CRM), the Event Message, and the Data Acquisition Message (DAQ). The CRM is a message sent by the ECU in response to a Command Receive Object (CRO). It can be a simple acknowledgement to the CRO or contain the actual requested data. An event message can be used by the slave ECU to signal a malfunctioning and possibly request that an error recovery procedure is started for the other ECUs. The format of a Command Return Message or an Event Message is shown in Fig. 9.29: byte 0 is the package ID (PID), byte 1 is the code for command return or error (ERR), byte 2 is the command counter (CTR) as received in the last CRO message, and the remaining bytes contain the data sent by the ECU in response to the master.

The Data Acquisition (DAQ) message is a specific type of DTO used to send measurement data to the tool, following the initialization phase. The Data Acquisition Message has a different format than the CRM or event messages, where byte 0 is the PID, and the other 7 bytes are the data field, as shown in Fig. 9.30. The PID in the DAQ message is used to specify the tables that are used to identify the location of the measurement data. Those tables are called Object Descriptor Tables (ODTs). They describe which data acquisition elements (i.e. variables) are contained in the remaining 7 data bytes of the message. Multiple ODTs may be needed to store all the requested measurement data, as the number of data bytes sent back in each ODT is limited to seven data bytes.

9.3.4 List of CCP Commands

Table 9.8 summarizes the list of supported commands. The 8-bit command code is represented by a hexadecimal number, followed by the timeout to acknowledgement, a flag that indicates whether the command is optional, and a short description of the command functionality. Please refer to [8] for the details of each command.

Table 9.8 List of CCP commands

Command	Code	ACK timeout (ms)	Optional	Description
CONNECT	0x01	25		Establish a logical connection to an ECU
GET_CCP_VERSION	0x1B	25		Get implemented Version of CCP
EXCHANGE_ID	0x17	25		Get the ECU identification
GET_SEED	0x12	25	Yes	Get the access protection code
UNLOCK	0x13	25	Yes	Unlock the access protection
SET_MTA	0x02	25		Set memory transfer address (MTA)
DNLOAD	0x03	25		Download ≤5 bytes into ECU memory
DNLOAD_6	0x23	25	Yes	Download 6 bytes into ECU memory
UPLOAD	0x04	25		Upload ≤5 bytes from ECU memory
SHORT_UP	0x0F	25	Yes	Upload ≤5 bytes from ECU memory (no MTA)
SELECT_CAL_PAGE	0x11	25	Yes	Select the active calibration page
GET_DAQ_SIZE	0x14	25	Yes*	Get the size of a DAQ list
SET_DAQ_PTR	0x15	25	Yes*	Set the ODT entry pointer
WRITE_DAQ	0x16	25	Yes*	Write an entry in a ODT
START_STOP	0x06	25	Yes*	Start /Stop data transmission
DISCONNECT	0x07	25		Finish a logical connection
SET_S_STATUS	0x0C	25	Yes	Set the ECU session status
GET_S_STATUS	0x0D	25	Yes	Get the ECU session status
BUILD_CHKSUM	0x0E	30,000	Yes	Calculate a memory checksum
CLEAR_MEMORY	0x10	30,000	Yes	Clear a memory range
PROGRAM	0x18	100	Yes	Download ≤5 program bytes
PROGRAM_6	0x22	100	Yes	Download 6 program bytes
MOVE	0x19	30,000	Yes	Move a memory range
TEST	0x05	25	Yes	Test for presence of a slave device
GET_ACTIVE_CAL_PAGE	0x09	25	Yes**	Get currently active calibration page
START_STOP_ALL	0x08	25	Yes	Start or stop all DAQ lists
DIAG_SERVICE	0x20	500	Yes	Diagnostic service
ACTION_SERVICE	0x21	5,000	Yes	Request action to the ECU

* Required if the ECU supports DAQ; ** Required if SELECT_CAL_PAGE is implemented

Table 9.9 List of packet IDs and their meaning

PID	Interpretation
0xFF	DTO contains a Command Return Message
0xFE	DTO contains an Event Message where CTR is don't care
$n \in [0, 0xFD]$	DTO contains a Data Acquisition Message corresponding to ODT n

Table 9.10 List of command return codes

Code	Description	ECU next state
0x00	Acknowledge/no error	
0x01	DAQ processor overload	
0x10	Command processor busy	None (wait until ACK or timeout)
0x11	DAQ processor busy	None (wait until ACK or timeout)
0x12	Internal timeout	None (wait until ACK or timeout)
0x18	Key request	None (embedded seed and key)
0x19	Session status request	None (embedded SET_S_STATUS)
0x20	Cold start request	Cold start
0x21	Cal. data init. request	Cal. data initialization
0x22	DAQ list init. request	DAQ list initialization
0x23	Code update request	(Cold start)
0x30	Unknown command	(Fault)
0x31	Command syntax	Fault
0x32	Parameter(s) out of range	Fault
0x33	Access denied	Fault
0x34	Overload	Fault
0x35	Access locked	Fault
0x36	Resource/function not available	Fault

9.3.5 List of Packet IDs and Command Return Codes

As summarized in Table 9.9, the PID field of a CRM has the value **0xFF**, and the same field has the value **0xFE** for the Event Message. The list of possible command return codes along with their short description and the ECU state following the command is described in Table 9.10.

9.4 TTCAN

TTCAN (Time-Triggered CAN) is a higher layer protocol defined on top of the CAN data-link layer standards to allow time-triggered operation on CAN. TTCAN has been standardized in ISO 11898-4 [9]. It provides support for time-triggered message transmissions, in addition to event-triggered communication in a standard CAN system. The motivation for the introduction of such a protocol, as in other time-triggered protocols, is to guarantee a deterministic communication pattern on the bus, in order to avoid latency jitter and improve communication efficiency.

9.4.1 *Motivation and Goal*

The bitwise arbitration at the medium access level in CAN is often considered as one of its most powerful features. Collisions among identifier bits are resolved by the logical AND semantics, thus if multiple messages are ready, the one with the lowest identifier (highest priority) is granted access for transmission. The transmission of all the other messages will be delayed (at least) until the beginning of the next arbitration phase. However, in standard CAN, there is no clock synchronization among the nodes in the network, and the ready time of messages on other nodes is random. Hence, messages experience a variable latency depending on the readiness of other messages in the network. In general, the lower the message priority, the higher the latency jitter.

Today, several innovative automotive control systems are being introduced in vehicles, e.g. x-by-wire systems, motor management, and sensor subnetworks. These systems require a deterministic behavior in the communication. In time-triggered communication any activity in the network including message transmission is determined by the progression of a globally synchronized time, which can provide a deterministic and predictable communication pattern on the bus to avoid latency jitters and improve communication efficiency [9]. TTCAN was developed by Robert Bosch GmbH on top of the CAN physical and data link layers, which was later standardized to support such communication guarantees.

9.4.2 *Synchronization Mechanisms in TTCAN*

As in any time-triggered architecture, TTCAN requires the definition and management of a globally synchronized time among the control units in the network. This is done by the periodic transmission of *reference messages* from a subset of network nodes denoted as *time masters*. TTCAN has two levels of synchronization: level 1 includes only the minimum functionality for a time-triggered operation, which includes some degree of fault-tolerance with respect to the availability of the time masters; level 2 also establishes a global network time base with high precision.

The reference message is periodically sent by the time master and can be easily recognized by its identifier. The time interval between two consecutive reference messages is the *basic cycle*. The basic cycle length is not necessarily constant, in order to accommodate messages with different periods. In TTCAN level 1, the reference message only contains one byte of control information, and the rest of the message (up to 7 bytes) can be used for data transfer. In level 2, the reference message holds information on the global time of the current time master in the first 4 bytes, and the remaining 4 bytes can be used for data transfer.

As the synchronization mechanism relies heavily on the availability of the reference message, it is necessary to provide fault tolerance for the functionality of the time master. A predefined list of TTCAN controllers (up to eight) is identified with a full rank among them. If a higher rank time master fails, all the other potential time masters must identify the failure and react within a short timeout. All candidate time masters start a recovery protocol by trying to send the reference message. Given that the identifier of a reference message from a higher rank node has higher priority, the message from the highest rank candidate master wins the arbitration. Correspondingly, the sender node automatically takes over as the new time master. At any time, a higher rank time master that resumes operation and its functionality can regain its role by sending its own reference message at the start of the next basic cycle.

The *Network Time Unit* (NTU) is the granularity of any timing information in the network, typically in the order of a CAN bit time. A consistent NTU among all nodes is established by a clock synchronization protocol that relates the information in the reference message with the local time and its rate of advancement, as defined by the local oscillator. The node-dependent oscillator circuit defines the local view of the system clock, which is constantly adjusted by the local Time Unit Ratio (TUR) to take care of the correct relation between the system clock and the NTU.

In TTCAN level 2, all nodes takes a snapshot of the local time values at the Start of Frame (SoF) bit of the reference message. After the successful reception of the reference message which contains the global time from the time master, the difference between the global time and local time is recorded as the local offset. During the next basic cycle the node adjusts its local time using the local offset to correct its offset with respect to the global time.

Due to the different clock speed of different nodes, an additional adjustment has to be made to guarantee the consistency between the local time and global time. This is done by the continuous update of TUR as follows. The length between two consecutive reference message SoF synchronization points is measured both globally (from the data values transmitted in the reference message) and locally (number of oscillator period in this interval). The quotient of these two lengths gives the actual TUR value. A precision of 1 μs for the global time synchronization algorithm is achieved by these protocols.

9.4.3 Scheduling of TTCAN Networks

The basic cycle starts at the synchronization point of the reference message. Besides the reference message itself, the basic cycle is divided into three types of windows, namely, the exclusive window, the arbitration window, and the free window. The *exclusive window* is dedicated to the transmission of a predefined message. If the system schedule is properly defined, there should be no conflict during the exclusive window. The exclusive window is typically used for the transmission of periodic messages. The *arbitration window* is used for messages that are sporadic or generated after an event (event-driven). In the arbitration window, TTCAN uses the bit-wise arbitration as in standard CAN to decide which message gets to be transmitted next. The *free window* is reserved for future extensions during design time.

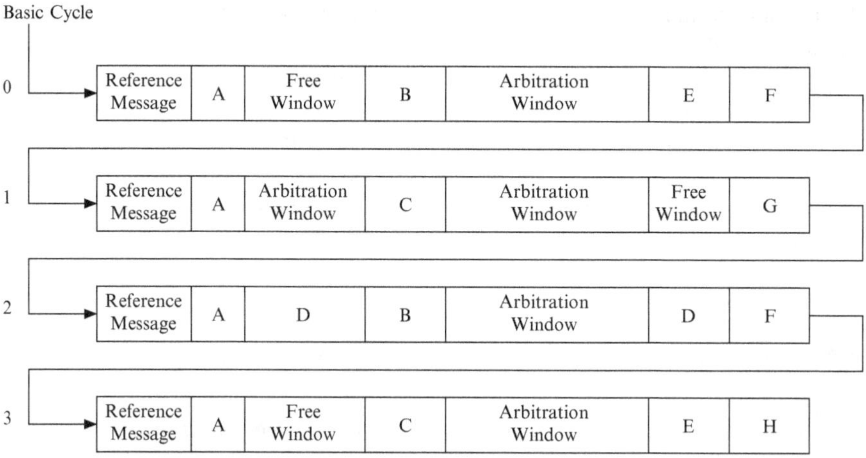

Fig. 9.31 An example of a TTCAN system matrix

To provide flexibility to the designers, an exclusive time window can be repeated more than once in one basic cycle. Also, periodic messages may have different periods due to the requirements of different feedback control loops. Thus, several basic cycles are combined to build the system matrix. An example is shown in Fig. 9.31, where the messages labeled A to H are periodic messages transmitted in exclusive time windows.

The scheduling information in the *system matrix* should be properly specified off-line. This design activity includes the selection of the following parameters:

- The number of basic cycles in the system matrix
- The number of time windows within the basic cycle
- The messages to be transmitted in each time window
- The duration of each time window

Message scheduling must comply with the pre-defined message periods and transmission times. The average message period can be calculated by the duration of the system matrix divided by the number of times a message appears in it. The length of the exclusive window can be computed once the mapping of each periodic message to its window is defined. The duration of the arbitration window needs to be large enough to accommodate all the event-driven messages assigned to it. Also, a latest transmission time *Tx_Enable* is defined at design time for the arbitration window, to guarantee that no sporadic message will spill into the following periodic time window. At runtime, no message is allowed to start transmission after Tx_Enable passes in the arbitration window. A careful design of the system scheduling combined with the use of exclusive and arbitration windows gives the advantage of both time-triggered and event-triggered communication.

All nodes should store the local scheduling information prior to the start of the operations. The node-specific knowledge must be kept at a minimum to minimize the memory utilization. It includes the messages to be sent and received in the exclusive windows, as well as the messages to be sent in the arbitration window. All the functionalities in TTCAN level 1 can be implemented in software on top of standard CAN devices. This requires practically no change to the commercial CAN hardware used in the network. Level 2 requires the additional ability to time-stamp the Start of Frame-bit of the reference message and send the time-stamp within the very same message (a span of several microseconds). Hardware support is therefore required in probably all cases.

9.4.4 Reliability of TTCAN and Additional Notes

Of course, in the transmission of periodic time-triggered messages in the synchronous window, the CAN peripheral should be configured to skip the automatic retransmission of a message in case of a network error. This feature is available in some but not all CAN controllers.

Also, errors may affect the operation of event-driven messages. In this case retransmissions after errors might be enabled, but the CAN driver should pay attention that messages are never transmitted past the *Tx_Enable* time bound.

As in other time-triggered networks, TTCAN is particularly sensitive to possible timing faults by nodes that send messages outside their windows, possibly spoiling the transmission of messages in other nodes. This can be seen as a special case of a node behaving as a babbling idiot. The safest option in this case would be the availability of bus guardians, allowing node access to the network only in the windows in which it is allowed. Unfortunately, despite proposals from the research community [17, 31, 35], bus guardians for TTCAN are not commercially available.

Finally, with respect to the vulnerabilities reported in Chap. 6, it is important to note how in TTCAN, because of the disabled retransmissions, any error on the last but one bit that is not consistently detected by all network nodes will result in an inconsistent omission (somewhat a more serious problem than the inconsistent duplicate message in the regular CAN case).

Symbols

1. m_i: the i-th message
2. $M_{i,j}$: the j-th instance of message m_i
3. μ_i: the length of message m_i in bytes
4. s_i: the smallest multiple of 8 bits that is larger than the actual payload of message m_i
5. id_i: CAN identifier of message m_i
6. P_i: priority of message m_i (id in Chap. 3)
7. N_i: the index of node transmitting message m_i
8. Υ_i^{src}: index of node transmitting the message m_i
9. Υ_i: node i
10. T_i: period or minimum inter-arrival time of message m_i
11. J_i: queuing jitter (also known as arrival jitter) of message m_i
12. D_i: deadline of message m_i
13. $A_{i,j}$: arrival time of message instance $M_{i,j}$
14. $Q_{i,j}$: queuing time of $M_{i,j}$
15. $F_{i,j}$: finish time of message instance $M_{i,j}$
16. $S_{i,j}$: start time of message instance $M_{i,j}$
17. e_i or $\mathcal{E}_i$: transmission time of message m_i on a CAN bus
18. w_i: queuing delay of message m_i
19. $w_i^{(k)}$: value of w_i at the k-th iteration of solving the fixed point iteration
20. $w_i(q)$: queueing delay of the q-th instance in the busy period
21. R_i: worst case response time of message m_i
22. $R_i(q)$: worst case response time of q-th instance of message m_i in the busy period
23. $\mathcal{R}_{ij}$: response time of message instance M_{ij}
24. B_i: blocking delay in w_i due to a lower priority message that is transmitted when m_i is transmitted
25. I_i: interference delay in w_i due to higher priority messages that are transmitted on the bus before m_i is transmitted
26. $hp(i)$: set of messages with priority higher than message m_i

M. Di Natale et al., *Understanding and Using the Controller Area Network Communication Protocol: Theory and Practice*, DOI 10.1007/978-1-4614-0314-2,

27. $hp(i)_k$: $\{\tau_j(m_j)|P_j < P_i \cap \Upsilon_j^{src} = \Upsilon_k\}$, the set of tasks (messages) with priority higher than P from ECU Υ_k
28. Br: bit rate of transmitting CAN bus
29. τ_{bit}: time of transmission for one bit on the network
30. p: number of protocol bits in a frame
31. t: a value on the time line
32. t_s: start time of busy period for a message
33. t_e: end time of busy period for a message
34. $\mathcal{O}_i$: initial phase of message m_i
35. $\mathcal{O}_{ij}$: relative offset of two messages m_i and m_j
36. Ψ_i: local phase of message m_i
37. $\mathcal{O}_{lk}$: relative phase between ECU_l and ECU_k
38. $O_{\Upsilon_i,\Upsilon_j}$: clock difference between ECUs Υ_i and Υ_j
39. T_p: period of polling task
40. τ_p: polling task
41. H: hyperperiod
42. $f_{\mathcal{W}_t^P}$: *pdf* or *pmf* of random variable $\mathcal{W}_t^P$
43. $f_{(\mathcal{W}_t^P,\mathcal{X}_t)}$: joint *pdf* or *pmf* of random variable $\mathcal{W}_t^P$ and $\mathcal{X}_t$
44. G_k^P: P-level backlog at the beginning of the k-th hyperperiod, i.e. $W_{(k-1)H}^P$
45. H_k^i: lcm of the periods of the messages in $hp(P_i)_{\Upsilon_k}$
46. W_t^P: P-level backlog at time t
47. U: utilization
48. U_i: utilization of message m_i
49. $V_t^{i,j}$: event $\mathcal{Q}_{i,j} > t$
50. $\overline{V}_t^{i,j}$: the complement event of $V_t^{i,j}$, i.e. event $\mathcal{Q}_{i,j} \leq t$
51. $W_t^{M_{i,j}}$: backlog of message instance $M_{i,j}$ at time t
52. X_t: queuing pattern at time t
53. $lp(P)$: $\{\tau_i(m_i)|P_i > P\}$, the set of lower priority messages
54. $B_{i,j}$: blocking time of message instance $M_{i,j}$
55. For a random variable, e.g., $\mathcal{E}_i$, we denote its average value as $\bar{E}_i$, its minimum and maximum value as $E_i^{\min}$ and $E_i^{\max}$.
56. τ: granularity
57. $I(P,t)$: P-level message instances at time t
58. $S(P,t)$: P-level event space at time t
59. $\overline{X}_{t(i,j)}$: the complement queuing pattern of X_t with respect to $M_{i,j}$
60. x_i^{off}: the minimum message response time calculated as the sum of the message transmission time and the queuing delay from local higher priority messages using the expectation-maximization (EM) algorithm
61. $(a_i, b_i, y_i^D, y_i^\Gamma)$: parameters for the expectation-maximization (EM) algorithm (refer to Sect. 5.2.2)
62. $(x_{i,k}^{\text{off}}, y_{i,k}, a_{i,k}, b_{i,k}, y_{i,k}^D, y_{i,k}^\Gamma)$: six characteristic parameters for degenerate and offsetted gamma distributions (refer to Sect. 5.2.2)
63. T_i^{hr}, the minimum greatest common divisor (*gcd*) of the higher priority messages for remote nodes
64. $U_i^r, U_i^{hr}, Q_{i,j}$, and Q^{hr} the predictor parameters for regression analysis (refer to Sect. 5.3.2)

References

1. AUTOSAR (AUTomotive Open System ARchitecture) consortium web page. *http://www.autosar.org*.
2. ISO/DIS 26262-1 road vehicles – functional safety. *Available from www.iso.org*.
3. The Ptolemy Project at the University of California at Berkeley. Project web page: *http://ptolemy.eecs.berkeley.edu/ptolemyII/index.htm*.
4. Vector CANbedded OEM-Specific Embedded Software Components for CAN Communication in Motor Vehicles. Product web page *http://www.vector.com/vi_canbedded_en.html*.
5. Iso 11898-1. road vehicles - interchange of digital information - controller area network (can) for high-speed communication. *ISO Standard-11898, International Standards Organisation (ISO)*, November 1993.
6. Road vehicles - interchange of digital information - controller area network (CAN) for high-speed communication. *ISO 11898*, 1993.
7. Low-speed serial data communication part 2: Low-speed controller area network (CAN). *ISO 11519-2*, 1994.
8. Ccp can calibration protocol, version 2.1. *ASAP Standard*, 1999.
9. Road vehicles-controller area network (can) part 4: Time-triggered communication. *ISO/WD11898-4*, December 2000.
10. OSEK/VDX Communication Specification Version 3.0.3. Available at *http://www.osek-vdx.org*, July 2004
11. Vector canalyzer v7.2. Product web page: *http://www.vector.com/vi_canalyzer_en.html*, 2010.
12. The candb++ editor. *http://www.vector.com/vi_candb_en.html*, 2011.
13. Preevision e/e architecture tool. *https://www.vector.com/vi_preevision_en.html*, 2011.
14. ATMEL. *AT89C51CC03 Datasheet, Enhanced 8-bit MCU with CAN Controller and Flash Memory*. web page: http://www.atmel.com/dyn/resources/prod_documents/doc4182.pdf
15. M. Barranco, G. R. Navas, J. Proenza, and L. Almeida. Cancentrate: An active star topology for can networks. *Proceedings of the WFCS Workshop*, 2004.
16. R. Bosch. CAN specification, version 2.0. Stuttgart, 1991.
17. I. Broster and A. Burns. An analyzable bus-guardian for event-triggered communication. *Proceedings of the IEEE Real-Time Systems Symposium*, 2003.
18. I. Broster and A. Burns. Random arrivals in fixed priority analysis. In *1st International PARTES Workshop*, 2004.
19. L. Casparsson, Antal Rajnak, Ken Tindell, and P. Malmberg. Volcano revolution in on-board communications. *Volvo Technology Report*, 1998.
20. L. Cucu. Preliminary results for introducing dependent random variables in stochastic feasibility analysis on can. In *7th IEEE International WFCS Workshop, WIP session*, May 2004.

M. Di Natale et al., *Understanding and Using the Controller Area Network Communication Protocol: Theory and Practice*, DOI 10.1007/978-1-4614-0314-2,

21. Robert I. Davis, Alan Burns, Reinder J. Bril, and Johan J. Lukkien. Controller area network (can) schedulability analysis: Refuted, revisited and revised. *Real Time Systems Journal*, 35(3):239–272, 2007.
22. Robert I. Davis, Alan Burns, Reinder J. Bril, and Johan J. Lukkien. Controller area network (can) schedulability analysis: Refuted, revisited and revised. *Real-Time Systems*, 35(3):239–272, 2007.
23. José Luis Díaz, Daniel F. García, Kanghee Kim, Chang-Gun Lee, Lucia Lo Bello, José María López, Sang Lyul Min, and Orazio Mirabella. Stochastic analysis of periodic real-time systems. In *Proceedings of the 23rd IEEE Real-Time Systems Symposium (RTSS'02)*, pages 289–302, 2002.
24. J. Eker, J. W. Janneck, E. A. Lee, J. Liu, X. Liu, J. Ludvig, S. Neuendorffer, S. Sachs, and Y. Xiong. Taming heterogeneity—the ptolemy approach. *Proceedings of the IEEE*, 91(2), 2003.
25. Mohammad Farsi. *CANopen Implementation : Applications to Industrial Networks*. Research Studies Pre, May 2000.
26. J. Ferreira, A. Oliveira, P. Fonseca, and J. A. Fonseca. An experiment to assess bit error rate in can. In *Proceedings of the 3rd International Workshop of Real-Time Networks (RTN2004)*, pages 15–18, 2004.
27. Association for Standardisation of Automation and Measuring Systems (ASAM). Field bus exchange format, v4.0.0. *http://www.asam.net*, 2011.
28. Fujitsu. *MB90385 Series Datasheet*. Available from the Fujitsu web page: http://edevice.fujitsu.com/.
29. Mark K. Gardner. Probabilistic analysis and scheduling of critical soft real-time systems. *UIUC, Phd Thesis, Computer Science, University of Illinois at Urbana-Champaign*, 1999.
30. B. Gaujal and N. Navet. Fault confinement mechanisms on can: analysis and improvements. *IEEE Transaction on Vehicular Technology*, 54(3):1103–1113, 2005.
31. G. Buja, J. R. Pimentel, and A. Zuccollo. Overcoming babbling-idiot failures in the flexcan architecture: A simple bus-guardian. *Proceedings of the IEEE ETFA Conference*, 2005.
32. Infineon. *Infineon XC161CJ-16F, Peripheral Units User's Manual*. available from: http://www.keil.com/dd/docs/datashts/infineon/xc161_periph_um.pdf.
33. Intel. *8XC196Kx, 8XC196Jx, 87C196CA Microcontroller Family User's Manual*. Datasheet available from the web page: http://www.datasheetcatalog.org/datasheet/Intel/mXtwtqv.pdf.
34. G. Arroz, C. Almeida, J. Rufino, P. Verissimo and L. Rodrigues. Fault-tolerant broadcasts in can. In *28th International Symposium on Fault-Tolerant Computing Systems, Munich, Germany*, pages 150–159, June 1998.
35. J. Ferreira, L. Almeida, and J. Fonseca. Bus guardians for can: a taxonomy and a comparative study. *Proceedings of the WDAS Workshop*, 2005.
36. Jong Kim and Kang G. Shin. Execution time analysis of communicating tasks in distributed systems. *IEEE Transactions on Computers*, 45(5):572–579, May 1996.
37. Philip Koopman, Eushiuan Tran, and Geoff Hendrey. Toward middleware fault injection for automotive networks. In *28th International Symposium on Fault-Tolerant Computing Systems, Munich, Germany*, June 1998.
38. Bruce Davie Larry Peterson. Computer networks: A systems approach. In *Morgan Kaufmann, San Francisco*, 1996.
39. J. Lehoczky, L. Sha, and Y. Ding. The rate monotonic scheduling algorithm: Exact characterization and average case behavior. In *Proceedings of the IEEE Real-Time Systems Symposium*, December 1989.
40. John P. Lehoczky. Real-time queueing theory. In *Proceedings of the 17th IEEE Real-Time Systems Symposium (RTSS'96)*, pages 186–195, December 1996.
41. Wang Lei, Zhaohui Wu, and Mingde Zhao. Worst-case response time analysis for osek/vdx compliant real-time distributed control systems. In *Proceedings of the 28th Annual International Computer Software and Applications Conference (COMPSAC'04)*, pages 148–153, 2004.

42. C. L. Liu and J. W. Layland. Scheduling algorithms for multiprogramming in a hard-real-time environment. *Journal of the Association for Computing Machinery*, 20(1), 1973.
43. José María López, José Luis Díaz, Joaquín Entrialgo, and Daniel García. Stochastic analysis of real-time systems under preemptive priority-driven scheduling. *Real-Time Syst.*, 40(2):180–207, 2008.
44. Sorin Manolache. Schedulability analysis of real-time systems with stochastic task execution times. *PhD Thesis, Department of Computer and Information Science, IDA, Linkoping University*, 2002.
45. R. T. McLaughlin. Emc susceptibility testing of a can car. In *SAE Conference - SAE 932866*, 1993.
46. Antonio Meschi, Marco Di Natale, and Marco Spuri. Priority inversion at the network adapter when scheduling messages with earliest deadline techniques. In *Proceedings of Euromicro Conference on Real-Time Systems*, June 12-14 1996.
47. Microchip. *MCP2515 Datasheet, Stand Alone CAN Controller with SPI Interface*. Datasheet available at the web page: http://ww1.microchip.com/downloads/en/devicedoc/21801d.pdf.
48. Marco Di Natale. Evaluating message transmission times in controller area networks without buffer preemption. In *Brazilian Workshop on Real-Time Systems*, 2006.
49. N. Navet, Y. -Q. Song, and F. Simonot. Worst-case deadline failure probability in real-time applications distributed over controller area network. *Journal of Systems Architecture*, 46(7):607–617, 2000.
50. Thomas Nolte, Hans Hansson, and Christer Norström. Probabilistic worst-case response-time analysis for the controller area network. In *Proceedings of the 9th IEEE Real-Time and Embedded Technology and Applications Symposium (RTAS)*, pages 200–207, Washington, DC, USA, May 2003.
51. Thomas Nolte, Hans Hansson, Christer Norström, and Sasikumar Punnekkat. Using bit-stuffing distributions in can analysis. In *Proceedings of the IEEE/IEE Real-Time Embedded Systems Workshop (RTES)*, London, UK, December 2001.
52. Christian Keidel Olaf Pfeiffer, Andrew Aire. *Embedded Networking with CAN and CANopen*. Copperhill Media Corporation, April 2008.
53. J. C. Palencia and M. Gonzáles Harbour. Schedulability analysis for tasks with static and dynamic offsets. In *Proceedings of the 19th IEEE Real-Time Systems Symposium (RTSS'98)*, pages 26–37, December 1998.
54. R. Kaas Peter van Emde Boas and E. Zijlstra. Design and implementation of an efficient priority queue. *Mathematical Systems Theory*, 10:99–127, 1977.
55. Philips. *P8xC592; 8-bit microcontroller with on-chip CAN Datasheet*. Datasheet available at the web page: http://www.nxp.com/documents/data_sheet/P8XC592.pdf.
56. J. Proenza, M. Barranco, and L. Almeida. Recancentrate: A replicated star topology for can networks. *Proceedings of the ETFA Conference*, 2005.
57. Kai Richter, Dirk Ziegenbein, Marek Jersak, and Rolf Ernst. Model composition for scheduling analysis in platform design. In *Proceedings 39th Design Automation Conference (DAC 2002)*, June 2002.
58. K. Tindell, A. Burns, and A. Wellings. An extendible approach for analysing fixed priority hard real-time tasks. *Journal of Real Time Systems*, 6(2):133–151, Mar 1994.
59. Ken Tindell and Alan Burns. Guaranteeing message latencies on control area network (can). *Proceedings of the 1st International CAN Conference*, 1994.
60. Ken Tindell, Hans Hansson, and Andy J. Wellings. Analysing real-time communications: Controller area network (can). *Proceedings of the 15th IEEE Real-Time Systems Symposium (RTSS'94)*, 3(8):259–263, December 1994.
61. Eushiuan Tran. Multi-bit error vulnerabilities in the controller area network protocol. In *PhD Thesis, Carnegie Mellon University*, May 1999.
62. Wilfried Voss. *A Comprehensible Guide to J1939*. Copperhill Technologies Corporation, 2008.
63. Haibo Zeng. *Probabilistic Timing Analysis of Distributed Real-time Automotive Systems*. PhD thesis, EECS Department, University of California, Berkeley, Dec 2008.

64. Haibo Zeng, Marco Di Natale, Paolo Giusto, and Alberto Sangiovanni-Vincentelli. Statistical analysis of controller area network message response times. In *SIES'09: Proceedings of the IEEE International Symposium on Industrial Embedded Systems*, pages 1–10, July 2009.
65. Haibo Zeng, Marco Di Natale, Paolo Giusto, and Alberto Sangiovanni-Vincentelli. Stochastic analysis of can-based real-time automotive systems. *IEEE Transactions on Industrial Informatics*, 5(4):388–401, November 2009.

Index

M. Di Natale et al., *Understanding and Using the Controller Area Network Communication Protocol: Theory and Practice*, DOI 10.1007/978-1-4614-0314-2,

CPSIA information can be obtained
at www.ICGtesting.com
Printed in the USA
LVHW06*1150030618
579368LV00001B/15/P